Agents and Robots for Reliable Engineered Autonomy

Agents and Robots for Reliable Engineered Autonomy

Editors

Rafael C. Cardoso
Angelo Ferrando
Daniela Briola
Claudio Menghi
Tobias Ahlbrecht

MDPI • Basel • Beijing • Wuhan • Barcelona • Belgrade • Manchester • Tokyo • Cluj • Tianjin

Editors

Rafael C. Cardoso
Department of Computer Science
The University of Manchester
Manchester
United Kingdom

Angelo Ferrando
Department of Computer Science, Bioengineering, Robotics and Systems Engineering (DIBRIS)
University of Genova
Genova
Italy

Daniela Briola
Department of Informatics, Systems and Communication (DISCO)
University of Milano Bicocca
Milan
Italy

Claudio Menghi
Interdisciplinary Centre for Security, Reliability and Trust
University of Luxembourg
Luxembourg
Luxembourg

Tobias Ahlbrecht
Department of Informatics
Clausthal University of Technology
Clausthal-Zellerfeld
Germany

Editorial Office
MDPI
St. Alban-Anlage 66
4052 Basel, Switzerland

This is a reprint of articles from the Special Issue published online in the open access journal *Journal of Sensor and Actuator Networks* (ISSN 2224-2708) (available at: http://www.mdpi.com).

For citation purposes, cite each article independently as indicated on the article page online and as indicated below:

LastName, A.A.; LastName, B.B.; LastName, C.C. Article Title. *Journal Name* **Year**, *Volume Number*, Page Range.

ISBN 978-3-0365-1859-6 (Hbk)
ISBN 978-3-0365-1860-2 (PDF)

Contents

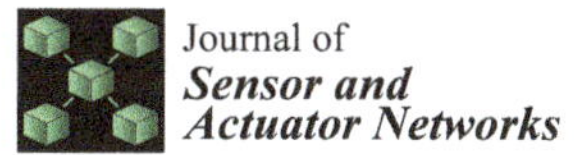

Journal of
*Sensor and
Actuator Networks*

Editorial

Special Issue: Agents and Robots for Reliable Engineered Autonomy

Rafael C. Cardoso [1,*], **Angelo Ferrando** [2,*], **Daniela Briola** [3,*], **Claudio Menghi** [4,*] and **Tobias Ahlbrecht** [5,*]

[1] Department of Computer Science, The University of Manchester, Manchester M13 9PL, UK
[2] Department of Computer Science, Bioengineering, Robotics and Systems Engineering (DIBRIS), University of Genova, 16145 Genova, Italy
[3] Department of Informatics, Systems and Communication (DISCO), University of Milano Bicocca, 20126 Milan, Italy
[4] Interdisciplinary Centre for Security, Reliability and Trust, University of Luxembourg, L-4365 Luxembourg, Luxembourg
[5] Department of Informatics, Clausthal University of Technology, 38678 Clausthal-Zellerfeld, Germany
[*] Correspondence: rafael.cardoso@manchester.ac.uk (R.C.C.); angelo.ferrando@dibris.unige.it (A.F.); daniela.briola@unimib.it (D.B.); claudio.menghi@uni.lu (C.M.); tobias.ahlbrecht@tu-clausthal.de (T.A.)

Citation: Cardoso, R.C.; Ferrando, A.; Briola, D.; Menghi, C.; Ahlbrecht, T. Special Issue: Agents and Robots for Reliable Engineered Autonomy. *J. Sens. Actuator Netw.* **2021**, *10*, 47. https://doi.org/10.3390/jsan10030047

Received: 2 July 2021
Accepted: 6 July 2021
Published: 13 July 2021

Publisher's Note: MDPI stays neutral with regard to jurisdictional claims in published maps and institutional affiliations.

The study of autonomous agents is a well-established area that has been researched for decades, both from a design and implementation viewpoint. Nonetheless, the application of agents in real-world scenarios is largely adopted when logical distribution is needed, but it is still limited when physical distribution is necessary. In parallel, robots are no longer used only in industrial applications but are instead applied in an increasing number of domains, ranging from robotic assistants to search and rescue. Robots in these applications often benefit from (or require) some level of (semi or full) autonomy. Thus, multi-agent solutions can be exploited in robotic scenarios, considering their strong similarity both in terms of logical distribution and interaction among autonomous entities.

The autonomous behavior responsible for decision making should (ideally) be verifiable since these systems are expensive to produce and are often deployed in safety-critical situations. Thus, verification and validation are important and necessary steps toward providing assurances about the reliability of autonomy in these systems. Likewise, software engineering techniques are an integral part of development in order to make sure that the systems meet requirements.

This Special Issue brings together researchers from the autonomous agents, software engineering and robotics communities, as combining knowledge from these three research areas may lead to innovative approaches that solve complex problems related with the verification and validation of autonomous robotic systems. We (the Special Issue editors) have written a perspective paper (peer-reviewed by members of the editorial board of the journal) that is included in our Special Issue [1]; in the perspective, we give an overview of recent research trends for researchers that aim at working in the intersection of these research areas.

This Special Issue was created based on the topics discussed at the First Workshop on Agents and Robots for reliable Engineered Autonomy (https://area2020.github.io/ accessed on 30 June 2021) (AREA 2020). One of the accepted papers in the workshop was invited to extend their work and underwent additional peer-review evaluation before being accepted for publication [2]. Paper [2] contains details about an architecture for linking robots with autonomous agents applied to a case study of campus mail delivery.

The other three remaining papers, refs. [3–5], were all new submissions. Paper [3] introduces a novel formal verification approach by combining the use of two well-known model checkers to verify the decision making in self-driving vehicles; it evaluates the resulting hybrid technique on a robotic simulator with a rational agent performing the decision making. Paper [4] is also applied to the autonomous automotive domain using agent-based control, but it is focused on the traffic rules that govern road junctions; it applies two different model checkers to assess the behavior of the agent at two different levels (design and development). Paper [5] provides an interpretation of the multi-agent systems as part of the aggregate programming paradigm to support the programming of collective autonomous behavior. Table 1 shows an overview of the main areas discussed in each paper of the Special Issue.

Table 1. Distribution of the papers across the main areas of the Special Issue.

	Autonomous Agents	Robotics	Verification and Validation	Software Engineering
Paper [1]	✓	✓	✓	✓
Paper [2]	✓	✓		
Paper [3]	✓	✓	✓	
Paper [4]	✓		✓	
Paper [5]	✓			✓

We hope that the different research communities that are represented in our Special Issue will improve their collaboration efforts in the future so that the best proposals from these different areas can be combined to create new and exciting solutions and tools to be exploited both in academia and in industry.

Funding: This work received no external funding.

Acknowledgments: First and foremost, we would like to thank all authors that have submitted their work to this Special Issue as well as the anonymous reviewers that made many comments and suggestions leading to improved versions of the submissions. We would also like to thank the support of everyone in the Journal of Sensor and Actuator Networks editorial office—in particular, the managing editor Louise Liu—and the academics from the editorial board that contributed to the editing process.

Conflicts of Interest: The authors declare no conflict of interest.

References

1. Cardoso, R.C.; Ferrando, A.; Briola, D.; Menghi, C.; Ahlbrecht, T. Agents and Robots for Reliable Engineered Autonomy: A Perspective from the Organisers of AREA 2020. *J. Sens. Actuator Netw.* **2021**, *10*, 33. [CrossRef]
2. Onyedinma, C.; Gavigan, P.; Esfandiari, B. Toward Campus Mail Delivery Using BDI. *J. Sens. Actuator Netw.* **2020**, *9*, 56. [CrossRef]
3. Al-Nuaimi, M.; Wibowo, S.; Qu, H.; Aitken, J.; Veres, S. Hybrid Verification Technique for Decision-Making of Self-Driving Vehicles. *J. Sens. Actuator Netw.* **2021**, *10*, 42. [CrossRef]
4. Alves, G.V.; Dennis, L.; Fisher, M. A Double-Level Model Checking Approach for an Agent-Based Autonomous Vehicle and Road Junction Regulations. *J. Sens. Actuator Netw.* **2021**, *10*, 41. [CrossRef]
5. Casadei, R.; Aguzzi, G.; Viroli, M. A Programming Approach to Collective Autonomy. *J. Sens. Actuator Netw.* **2021**, *10*, 27. [CrossRef]

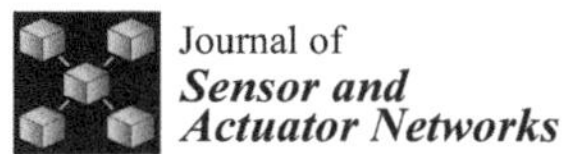

Journal of
*Sensor and
Actuator Networks*

Perspective

Agents and Robots for Reliable Engineered Autonomy: A Perspective from the Organisers of AREA 2020

Rafael C. Cardoso [1,*], Angelo Ferrando [2,*], Daniela Briola [3,*], Claudio Menghi [4,*] and Tobias Ahlbrecht [5,*]

[1] Department of Computer Science, The University of Manchester, Manchester M13 9PL, UK
[2] Department of Computer Science, Bioengineering, Robotics and Systems Engineering (DIBRIS), University of Genova, 16145 Genova, Italy
[3] Department of Informatics, Systems and Communication (DISCO), University of Milano Bicocca, 20126 Milan, Italy
[4] Interdisciplinary Centre for Security, Reliability and Trust, University of Luxembourg, L-4365 Luxembourg, Luxembourg
[5] Department of Informatics, Clausthal University of Technology, 38678 Clausthal-Zellerfeld, Germany
[*] Correspondence: rafael.cardoso@manchester.ac.uk (R.C.C.); angelo.ferrando@dibris.unige.it (A.F.); daniela.briola@unimib.it (D.B.); claudio.menghi@uni.lu (C.M.); tobias.ahlbrecht@tu-clausthal.de (T.A.)

Abstract: Multi-agent systems, robotics and software engineering are large and active research areas with many applications in academia and industry. The First Workshop on Agents and Robots for reliable Engineered Autonomy (AREA), organised the first time in 2020, aims at encouraging cross-disciplinary collaborations and exchange of ideas among researchers working in these research areas. This paper presents a *perspective* of the organisers that aims at highlighting the latest research trends, future directions, challenges, and open problems. It also includes feedback from the discussions held during the AREA workshop. The goal of this perspective is to provide a high-level view of current research trends for researchers that aim at working in the intersection of these research areas.

Keywords: multi-agent systems; robotics; software engineering; verification and validation; human–agent interaction

check for
updates

Citation: Cardoso, R.C.; Ferrando, A.; Briola, D.; Menghi, C.; Ahlbrecht, T. Agents and Robots for Reliable Engineered Autonomy: A Perspective from the Organisers of AREA 2020. *J. Sens. Actuator Netw.* **2021**, *10*, 33. https://doi.org/10.3390/jsan10020033

Academic Editor: Thomas Newe

Received: 15 March 2021
Accepted: 13 May 2021
Published: 14 May 2021

Publisher's Note: MDPI stays neutral with regard to jurisdictional claims in published maps and institutional affiliations.

1. Introduction

The robotics market is dramatically changing. Robots are more and more used to replace humans in their activities. For example, robots can be used in emergency search and rescue scenarios to reduce risks for humans rescuers. To operate in unpredictable environments, robots often need to be autonomous. Autonomous robots can perform their tasks with a high degree of autonomy without any human supervision. In addition, robots need to operate in a *reliable* manner to avoid failures that can have catastrophic consequences and lead to the loss of human life. The design of such robotic applications is complex since it requires engineers to consider different requirements related to different research domains.

The design of robotic applications requires *multi-agent solutions*. Robots are no longer only used in industrial applications, where robots operate in highly controllable and predictable environments. They are also used in an increasing number of domains, where the environment is often unpredictable and agents can have unexpected behaviours. For example, in an emergency search and rescue scenario the environment in which the robots operate is unpredictable: the structure of the buildings where robots are deployed may not be known in advance, and humans can have unpredictable reactions in emergencies. Robots in these applications often benefit from (or require) some level (semi or full) of autonomy. In addition, the missions the robots need to achieve are more and more complex and require multiple robots, with different capabilities, to collaborate. Thus, multi-agent solutions are required.

J. Sens. Actuator Netw. **2021**, *10*, 33. https://doi.org/10.3390/jsan10020033   https://www.mdpi.com/journal/jsan

Verification and Validation (V&V) aims at checking whether software behaves as expected. The distributed and autonomous nature of multi-agent systems poses a novel set of challenges for V&V. For example, the autonomous behaviour responsible for decision-making should (ideally) be extensively verified since these systems are expensive to produce and are often deployed in safety-critical situations. However, the autonomous behaviour of these systems is often unpredictable: it depends on the environmental conditions in which the system operates, and often changes at runtime. Thus, autonomous robotic systems are introducing a novel set of challenges that need to be addressed by novel V&V techniques.

Software Engineering (SE) refers to a branch of computer since that aims at supporting rigorous software development. Engineering multi-agent systems requires systematic and rigorous techniques that allow developing systems that meet their requirements. Multi-agent systems are instances of complex distributed systems. They require engineers to define the software architecture, to design the agent behaviours (e.g., through models), the protocols (if any) that agents should use to communicate, and how and when agents need to collaborate. Selecting and defining all of these components is often a difficult and complex activity, since it requires cross-disciplinary skills, and knowledge of multi-agent solutions, and the verification and validation to be used to support software design.

Finally, since multi-agent systems usually need to interact with people, engineers need to consider *human–agent interactions* as a key feature during the software design. Human–robot interactions are complex to be designed. Humans can (negatively) affect the behaviour of the robots, especially when humans and robots are collaborating for achieving certain goals. Human behaviour can trigger unexpected software reactions. Software components must be designed to adapt and modify their behaviours and to effectively support unexpected human actions.

Therefore, the design of complex robotic applications requires combining solutions coming from different research areas: multi-agent systems, verification and validation, software engineering, and human–agent interaction. This paper presents a *perspective* from the organisers of the first workshop on Agents and Robots for reliable Engineered Autonomy (https://area2020.github.io/ accessed: 14 May 2021) (AREA 2020). The goal of this workshop was to attract researchers from these areas, to support the exchange of ideas, and the cross-fertilisation and collaboration among the different research communities. This perspective presents some of the latest research trends and promising solutions in each of these areas. It is based on the research experience of the authors, and some of the discussions held during the AREA workshop. As such, it does not aim to be a complete and detailed review of the work done in these research areas, but it aims to be an initial read for researchers that aim at working in the intersection of these areas based on a speculative view of the organisers of AREA 2020.

This perspective paper is organised as follows. Section 2 concerns *multi-agent programming*. It discusses the use of programming languages designed for multi-agent systems for developing decision-making in robots, listing some of the tools for agents and robots individually and how these have been combined by the community. Section 3 concerns *verification and validation*. It presents works performed in the verification and validation with a special interest on multi-agent systems and robotic applications. Section 4 concerns *software engineering*. It describes research work in SE with a special interest in multi-agent systems. Section 5 concerns *human–agent interactions*. It describes research works that concern the development of robotic applications that consider human–agent interactions in the software design. Finally, Section 6 concludes our perspective. It summarises and discusses the challenges identified for each of the research areas we considered, and links the findings of the different sections.

2. Multi-Agent Programming

In this section, we will present some techniques that support the use of agent programming for the development of robotic applications. We will then present some challenges that prevent the effective use of agent programming for developing robotic systems.

One of the most popular models for agent programming languages is the Belief–Desire–Intention (BDI) model [1,2]. In the BDI model, there are three major concerns to model the agent behaviours: *beliefs*—knowledge that the agent has about the world, itself, and other agents; *desires*—goals that the agent wants to achieve; and *intentions*—recipes on how to achieve goals. Some examples of BDI agent programming languages include Jason [3] (used in many of the applications we cite in our perspective paper), JaCaMo [4,5] (Jason combined with two other technologies to provide first-class support for programming agents, environments, and organisations), and GWENDOLEN [6] (a more bare-bones agent language made specifically to support formal verification of agent programs).

The Robot Operating System (ROS) [7] is the de facto standard for the development of software for robotic applications. ROS supports the development of robotic applications through ROS nodes. An ROS node is a process that performs some computation. An ROS application is made by several nodes (representing components and subsystems of the robot) that communicate with one another following the publisher/subscriber model. The main advantage of ROS is its interoperability with different robot manufacturers and models. Since robot manufacturers provide support for executing the ROS, developers do not have to learn the firmware of each robot model.

An approach combining JaCaMo (in particular, the agent and environment layers) with ROS is presented in [8]. Their approach uses environment artifacts to implement ROS nodes and to provide actions for agents to publish and subscribe to nodes. Since environment artifacts in JaCaMo are implemented in Java, the approach makes use of the *rosjava* package (http://wiki.ros.org/rosjava accessed: 14 May 2021) (an implementation of some of ROS core features in Java) to make the artifacts able to interact with ROS. However, *rosjava* is not directly maintained by the ROS community. Therefore, it requires additional time to be supported after each new ROS release.

A Jason based integration with ROS is introduced in [9]. Their approach modifies Jason's reasoning cycle to support agents with the ability to subscribe and publish in topics from ROS nodes. Extra code for ROS (written in C++) is required for the integration as well.

In [10,11], Jason is linked with ROS through their SAVI ROS BDI architecture. This architecture is implemented in Java (using the *rosjava* package, and as such has the same disadvantage present in [8]) and mainly introduces a state synchronisation module that acts as a mediator between ROS and Jason by managing perceptions, incoming and outgoing messages, and actions being sent by the agent.

Differently from all of the above approaches, the authors in [12] propose an interface for programming autonomous agents in ROS that works without any changes to ROS or to any of the supported agent languages (Jason and GWENDOLEN). This is achieved through the *rosbridge* library [13], which allows code written in other languages (ROS has native support for C++ and Python only) to communicate with ROS topics through the WebSocket protocol.

Perspective of the Authors

As noted in recent agent programming reviews and surveys [14–16], there are still many open challenges that prevent agent programming languages to be used in the robotic domain. Some of these challenges include:

- the limited set of features provided by existing agent-based languages;
- immature methodologies and tools;
- no significant advantages for developers to change to agent programming since most applications can be implemented in more contemporary programming languages;

- the lack of quantitative and qualitative comparisons with other agent languages and other programming paradigms that guide developer in the selection of the most suitable language for their needs;
- the limited integration of agent-based technologies with other techniques, e.g., techniques coming from Artificial Intelligence (AI).

Additionally, there are other challenges that are more specifically related to the use of agent programming languages in robotic systems development.

One of these challenges is the limited support for high frequency data that coming from sensors, which are commonly used to represent "beliefs." When high-frequency data representing beliefs come from sensors, then the agent has to spend a large amount of time reasoning on these new perceptions. This reduces the performance of the robots in computing the new actions to be executed. A common solution to this problem is to use filters that limit the amount of data that is perceived, either by decreasing the frequency or limiting data based on its content and what would be interesting to the agent. However, these filters are often domain-specific and have to be tweaked based on the application.

Another challenge is the compatibility of agent languages with popular robotic frameworks such as ROS. An increasing number of languages are being extended by the community to work with ROS; however, these often modify either ROS or the agent language (or sometimes both) which can discourage new developers from using them.

An earlier survey [17] on agent languages for programming autonomous robots has identified four major challenges:

1. support for agent languages in robotic frameworks;
2. effectively managing sensor data into beliefs;
3. support for real-time reactivity;
4. synchronising robots while executing their plans.

As previously discussed, a significant amount of research has been conducted for addressing challenges 1 and 2. However, more work is needed to support additional agent-based languages, and more sophisticated and effective filtering techniques to manage sensor data. Less research has been conducted to address challenges 3 and 4, since these challenges do not always appear in robotic applications. For example, in a scenario in which a robot has to inspect a nuclear facility, the robot should be able to handle and adapt to failures. We believe that research on real-time reactivity of robots and the effective synchronisation of robots for executing their plans, on the one hand, will benefit from the additional support provided for existing agent-based languages and the more sophisticated and effective management, on the other hand, may highlight limitations of these frameworks and pave the way for further additional research.

3. Verification and Validation

As discussed in our introduction, reliability is very important in the design of robotic applications. However, multi-agent applications are posing a new set of challenges for the verification and validation (V&V) activities. In this section, we are considering both static verification techniques for MAS and dynamic verification techniques. Static formal verification techniques, such as Model Checking and Theorem Proving, usually check whether the system meets its requirements. Requirements are usually represented using formal specifications, a.k.a. properties, the system is usually represented using models. Dynamic formal verification techniques, such as Runtime Verification techniques, usually monitor the system execution and check whether observed behaviours meet the system requirements.

In this section, we describe the latest developments in the context of formal verification and validation of MAS (Section 3.1) and robotic systems (Section 3.2). Some of the works we present in this section were also discussed by a recent survey on formal verification applied to autonomous systems and robotic applications [18]. Then, we will argue (Section 3.3) that, as also argued by [19], robotic applications and autonomous systems pose a new set of challenges for formal verification and validation techniques.

3.1. Multi-Agent Systems

For MAS, V&V techniques usually check the behaviour of a set of agents collaborating or competing amongst themselves to achieve certain goals.

3.1.1. Model Checking

Model-checking exhaustively verifies a system against a formal property. It returns a Boolean verdict: *true* if the property is satisfied, and *false* and a counterexample, if it is not. Model-checking techniques are usually compute-intensive since the implemented procedures have a high temporal complexity because all the behaviours of the system need to be analysed for proving that the property is satisfied.

In [20], a model checker for verifying MAS, called MCMAS, is proposed. MCMAS supports temporal, epistemic and strategic properties. In its standard version, MCMAS requires to know the number of agents at design time. In [21], a parametric extension (MCMAS-P) to handle scenarios where the number of components cannot be determined at design time is presented, while, in [22], a more expressive extension (MCMAS-SL) is proposed to support strategy logic. MCMAS has been used in many different works for verifying MAS. In particular, in [23], where an analysis is carried out on the verification problem of synchronous perfect recall multi-agent systems with imperfect information. While the general problem is known to be undecidable, [23] shows that, if the agents' actions are public, then verification is decidable.

In [24], the authors propose a method for, and implement a working prototype of, an ethical extension to a rational agent governing an unmanned aircraft. Differently from [20], this work is focused on verifying BDI agents, defined using the agent language Ethan, an extension of GWENDOLEN. The resulting ethical agent is verified in AJPF [25], a model checker for agent programs. Differently from MCMAS, which verifies an abstract model of the MAS (i.e., a Concurrent Game Structure—CGS), AJPF verifies the source code of the MAS application. Furthermore, MCMAS assumes that properties are expressed using Alternating-Time Temporal Logic (ATL), which allows for reasoning on agents' strategies, while AJPF assumes that properties are expressed in Linear Temporal Logic (LTL) (enriched with epistemic logic operators).

In [26], the authors proposed the VERMILLION framework. VERMILLION targets a broad class of avionics systems that is amenable to analysis using formal methods. It extends the BDI model to incorporate learning, safety, determinism, and real-time response, and represents the abstract formal model using the Z language [27]. Compared to MCMAS, VERMILLION performs formal verification on an abstract model of the system, and not to the source code of the MAS. This requires engineers to build the abstract model of the MAS before running the verification framework.

Autonomous platoons are a typical example of MAS which are subject to extensive research. Formal verification of autonomous platoons has been considered for example in [28]. In this work, the authors proposed a reconfigurable multi-agent architecture to ensure the safety of the platoon, and specifically to guarantee a certain inter-vehicle distance among the different vehicles of the platoon. The authors proposed a model for the platoon that enables vehicles to join and leave the platoon, and verified whether this model ensures the satisfaction of a set of safety properties. Safety properties are formally verified using the Uppaal model checker [29]. Additionally, this work proposes to use the Webots simulator (https://github.com/cyberbotics/webots accessed: 14 May 2021) to simulate certain behaviors of the model.

Similarly, in [30], the authors applied formal verification to the model of the system and the actual implementation to ensure that autonomous decision-making agents in vehicle platoons never violate some safety requirements. In addition, in this work, the model checking procedure relies on the Uppaal model checker: the models of the agents are translated into timed automata, which are verified in Uppaal.

In [31], the authors present a new technique for model checking the logic of knowledge and commitments (CTLKC+). The proposed technique is fully-automatic and reduction-based. It reduces the problem of model checking CTLKC+ specifications to the problem of model checking an ARCTL [32] specifications. ARCTL is an existing logic that is supported by an existing model checker that relies on the NuSMV symbolic model checker [33].

3.1.2. Runtime Verification

There are various runtime verification techniques available in literature for MAS. In [34], the authors present a framework to verify agent interaction protocols at runtime. The formalism used in this work allows using variables to represent complex MAS behaviours. In [35], the authors extended their approach by supporting the usage of multiple monitors. Specifically, the global specification, which is used to represent the global protocol, is translated into partial decentralised specifications—one for each agent of the MAS. In [36,37], other works on runtime verification of agent interactions are proposed for the JADE platform. Specifically, in [36], the authors propose a framework called Multi-agent Runtime Verification (MARV). In this framework, requirements of MAS interaction during runtime are defined, such as availability and trustability. Differently from the other works, the requirements are expressed using natural language. The translation to a more formal representation is seen as a future work and not supported yet. Considering interactions in JADE, in [37], we may find a different approach which is partially obtained at runtime. In fact, the proposed method is performed in a semi-runtime way, where logs of messaging events are kept, and an algorithm for converting these logs to Time Petri Net as runtime program models is used.

When agents are dynamically adaptable, we may find application of runtime verification as presented in [38], where a runtime verification framework for dynamic adaptive MAS (DAMS-RV) based on an adaptive feedback loop is presented.

In [39], the authors propose a framework that combines model checking and runtime verification for analysing MAS. In this framework, the agents are first verified statically (using the AJPF [25] model checking), and, then, they are validated at runtime, through runtime verification using an extension of the work proposed in [34].

3.2. Robotic Applications

This section analyses works related to the formal V&V of robotic systems.

3.2.1. Model Checking

In [40], the authors propose an approach to formally verify an autonomous decision-making planner/scheduler system for an assisted living environment with the Care-O-bot robotic assistant. This is done by converting the robot house planner/scheduler rules into a multi-agent modelling language, i.e., Brahms model [41], and then, by translating this model into the PROMELA [42] language, which is then verified using the SPIN model checker [43]. Differently from the works we presented in the previous section, in this work, the model to be verified concerns scheduling rules used by an actual robot, rather than the reasoning process of an abstract agent.

In [44], the authors verified a formal model that describes mobile robot protocols operating in a discrete space is proposed. This formal model is then verified using the DiVinE model checker [45]. In [46], the authors propose an approach to verify real-time properties of ROS systems related to the communication between ROS nodes. Specifically, the authors analysed the source code of the Kobuki robot. Verification is performed by using the Uppaal model checker.

In [47], the authors analysed a collision avoidance protocol for multi-robot systems based on control barrier functions. The authors formally verified the properties of the collision avoidance framework. They showed that their controller formally guarantees collision free behaviour in heterogeneous multi-agent systems by applying slight changes to the desired controller via safety barrier constraints.

Finally, formal methods are also used to check whether the tasks of a robotic application can be scheduled with respect to a given hardware platform (e.g., [48,49]). For example, some of these works considered components specified in GenoM [50] (a middleware for robotic development similar to ROS) and automatically translate them into FIACRE [51], a formal language for timed systems.

3.2.2. Human–Robot Interaction

Building reliable software systems involving human–robot interactions poses significant challenges for formal verification. In [52], the authors propose a risk analysis methodology for collaborative robotic applications, which relies on well-known standards, and use formal verification techniques to automate the traditional risk analysis methods. In [53], the authors propose an innovative methodology, called SAFER-HRC, is presented. This methodology is centred around the logic language TRIO and the companion bounded satisfiability checker Zot [54], to assess the safety risks in a Human–Robot Collaboration (HRC) application.

3.2.3. Runtime Verification

Runtime Verification approaches for robotic applications are also discussed in the scientific literature. RobotRV [55] is a data-centred real-time verification approach for robotic systems. Within this approach, a domain-specific language named RoboticSpec is designed to specify the complex application scenario of the robot system.

Another runtime verification framework, called ROSMonitoring [56,57], allows the verification of ROS-based systems. In ROSMonitoring, runtime monitors are automatically synthesised from high-level specifications and used to verify formal properties against message passing amongst ROS nodes. The advantage of this framework is its being formalism-agnostic and portable. Indeed, the formal part of the monitors is decoupled and can be easily replaced.

3.2.4. Machine Learning

Machine learning is widely relevant for designing robotic applications. In [58], the authors consider the problem of formally verifying the safety of an autonomous robot equipped with a neural network controller that processes LiDAR images to produce control actions. The contributions are: (i) the definition of a framework for formally proving safety properties of autonomous robots equipped with LiDAR scanners; (ii) the notion of imaging-adapted partitions along with a polynomial-time algorithm for processing the workspace into such partitions; and (iii) a Satisfiability Modulo Convex (SMC)-based algorithm combined with an SMC-based pre-processing for computing finite abstractions of neural network controlled autonomous systems.

3.3. Perspective of the Authors

A lot of research was done on formal verification of MAS and Robotic applications. However, many challenges still need to be addressed. In the following, we will discuss two of these challenges.

- *scalability.* Many approaches suffer from scalability issues [59]. Researchers should find more efficient ways to verify the system under analysis especially when the number of agents and robots increases. Indeed, MAS and robotic applications are intrinsically distributed, and we expect the number of robots and agents of future robotic applications to increase over time. As previously mentioned, scalability issues are less relevant for dynamic verification approaches, such as runtime verification that only verify subsets of system executions. We believe that combining static and dynamic verification may be a valuable direction to address this challenge;
- *verification of ML components.* ML components are mode and more used in safety-critical scenarios (e.g., Robotic applications). However, the behaviours of machine learning components are usually not understandable by humans. Indeed, the behaviour of ML components is not defined a priori by humans, but ML components learn their behaviours from a set of training data. This poses the challenge of understanding if a ML component ensures the satisfaction of safety properties. While in

the past years many works have been proposed to enhance learning algorithms with formal methods, a lot of work needs to be done to make these approaches applicable in practice. For example, for MAS applications, some works have been proposed for single Reinforcement Learning agents, but few of them considered Multi-Agent Reinforcement Learning.

4. Software Engineering

This section provides a brief overview on some of the software engineering (SE) techniques that aim to support the development of reliable *multi-agent* and *robotic* applications. Researchers are extending, adapting, and creating new SE techniques to meet the needs of robotic applications. However, there are still many challenges that prevent the effective and efficient development of multi-agent systems and the community requires novel SE solutions. We introduce the overall main problems by giving an overview of them, citing some of the existing and known approaches and solutions, and discussing promising research trends and open problems.

Specifically, in this section, we discuss rigorous and systematic techniques that allow the specification of MAS requirements (Section 4.1), effective and efficient techniques that support testing MAS (Section 4.2), and simulation tools that enable reproducing the MAS and robotic behaviour with reasonable accuracy (Section 4.3). Finally, we will discuss our perspectives (Section 4.4).

4.1. Requirement Specification

The specification of the requirements of a multi-agent application is critical during software development. In MAS, requirements specification often concerns the definition of the task, also known as mission [60], what the application should achieve, and how to make the requirement executable by the MAS. Compared with conventional software, the presence of multiple agents makes the requirement specification phase more complex and error-prone since engineers need to precisely identify the different agents and identify the tasks they need to execute [61]. To support engineers in the specification of the requirements of the MAS application, several tools were proposed in the literature, such as natural languages, logic-based languages, pattern-based languages, domain-specific languages, and goal-modelling techniques.

In the following, we summarise some of the solutions proposed in these areas and evaluate how these solutions were used within the papers presented in the AREA 2020 workshop (https://area2020.github.io/ accessed: 14 May 2021), since they provide good examples of research covering different aspects of MASs. Specifically, Table 1 summarises the requirement specification technique used for each of the papers presented in the AREA 2020 workshop. Each row contains a requirement specification technique, i.e., natural languages, logic-based languages, pattern-based languages, domain-specific languages, and goal-modelling. Each column contains the reference to one of the papers presented in the AREA 2020 workshop. The cell at the intersection between a row of one requirement specification technique, and a column, of one paper presented in the AREA 2020 workshop, indicates the requirement specification technique used in that paper. For example, the marker at the intersection between the row labelled as "logic-based languages" and the column labelled with the reference [62] indicates that a logic-based language is the requirement specification technique used in [62].

J. Sens. Actuator Netw. **2021**, 10, 33

Table 1. Requirements specification technique used by the papers published in AREA 2020 ([63] is not reported in the table since it does not consider MAS requirements).

	[62]	[64]	[65]	[66]	[67]	[68]	[69]	[70]	[10]
Natural Language						✓			
Logic-based	✓	✓							
Pattern-based									
Domain-specific			✓	✓	✓			✓	✓
Goal-modelling									
Demonstrations							✓		

4.1.1. Natural Languages

In many contexts, requirements are initially expressed in natural language. This is a common case in many industrial applications (e.g., [71,72]). Natural languages offer significant benefits, they are easy to understand, and they support effective communication among different stakeholders. Several works considered the role of natural languages in the requirement specifications of multi-agent and robotic systems. For example, in [73], the authors proposed an approach to teach agents to communicate with humans in natural language. In [74], the authors analyse how to utilise and extend the Software Requirements Specifications model (IEEE Std 830-2009) to support the specification of requirements of multi-agent systems. In [75], the authors apply techniques of natural language processing for identifying the requirements and goals of multi-agent systems.

One paper presented at the AREA workshop assumed that the requirements of the MAS are specified using natural language. Specifically, in [68], the authors propose different types of interactions between an MAS and the final users who might benefit from communication-intensive, voice-based interactions.

4.1.2. Logic-Based Languages

Many researchers specify the requirements of the MAS in a logic-based language (e.g., [18,19,76–78]). Logic-based languages, such as LTL or CTL, assume that requirements are expressed using a set of atomic propositions that express relevant statements on the multi-agent system, combined with logical operators. One of the main advantages of using logic-based language is the availability of automated tools that support verification (e.g., model checking) and synthesis.

Two papers presented at the AREA workshop assumed that the requirements of the MAS are specified using logic-based languages. In [62], the authors use the TRIO [79] logical-language to specify the requirements of the MAS. TRIO is a first-order logical language. It provides temporal operators to constrain the values of some propositions at different time instants. In [64], the authors use the logical-based language provided by Uppaal [80] to specify the requirements of the MAS. Uppaal allows specifying missions through an extension of the CTL logic, which is a subset of TCTL. Specifically, it allows the specification of properties constraining a proposition to hold globally (resp. eventually) for every execution or querying whether there exists an execution such that a proposition holds globally (resp. eventually).

4.1.3. Pattern-Based Languages

Pattern-based languages are a common solution to solve recurrent problems of many domains, including robotics [60,81], cyber-physical systems [82], self-adaptive systems [83,84], machine learning [85], IoT [86], and multi-agent systems [84,87–89]. Existing design patterns in the field of multi-agent systems were classified in a recent survey [88].

By analysing the papers presented at the AREA workshop, we noticed that none of the published papers used pattern-based languages to specify the requirements of the multi-agent system.

4.1.4. Domain-Specific Languages

Several Domain-Specific Languages (DSL) for MAS were proposed in the literature (e.g., [90–92]). For MAS, DSLs usually provide constructs that enable engineers to model agents, the task they need to execute, and their interactions. Some of the recent DSLs for multi-agent systems are reported in the following.

In [93], the authors present a DSL for multi-robot application, based on the robotic mission specification patters [60]. In [94], the authors apply the DESIRE specification framework [95] on a case study on multi-agent systems. In [91], the authors propose a DSL for MAS. They also use the language and their graphical tool support for developing an MAS using a model-driven development approach.

Four papers presented at the AREA workshop proposed a DSL for the specification of a robot's core behaviours. In [65], the authors propose the use of Capability Analysis Tables (CATs). CATs provide a tabular representation that connects the inputs, outputs, and the behaviours of an MAS. Differently from other tools, e.g., logic-based languages, CATs are more understandable by non-expert users. We also considered the language used for specifying requirements by [66] as a domain-specific language. In this work, the authors assume that the requirement concerns reaching some specific states of the competence-aware systems used to model the MAS.

In [67], the authors use Jadescript [96] to specify the MAS and its requirements. Jadescript is a novel agent-oriented programming language compiled to Java. In [70], the authors consider the Planning Domain Definition Language (PDDL) [97] to specify the task to be performed by the MAS. PDDL is a DSL that is proposed to standardise automated planning languages. It enables the definition of the domain and the problem. The domain definition allows users to define predicates and operators (a.k.a. actions). The problem definition defines the objects of the problem instance, its initial state, and the goal. In [10], the authors use the Jason language to express the requirements of the MAS application. Jason is an agent-oriented programming language based on the BDI software model.

4.1.5. Goal-Modelling Techniques

Several goal-modelling techniques (e.g., Tropos [98], Gaia [99], Mobmas [100], INGE-NIAS [61]) support the development of multi-agent applications. These techniques enable users to identify the goals and the agents of the application. They also usually support the decomposition of goals into subgoals, and subgoals into tasks, and the assignment of tasks to agents.

By analysing the papers presented at the AREA workshop, we noticed that none of the published papers used goal-modelling techniques to specify the requirements of the multi-agent system.

4.1.6. Demonstrations

Many approaches use demonstrations to train the agents of a multi-agent application to perform their tasks [101–104]. Demonstration approaches usually require a human to demonstrate to the agent the task to be performed. Then, the agent learns and repeats the task.

One paper presented at the AREA workshop used demonstrations to specify the system goals. Specifically, in [69], the authors proposed a semi-supervised learning approach from demonstrations through program synthesis. Within this approach, a human operator specifies the goals by demonstrating to the agents how to perform the task. The MAS automatically infers high-level goals from the demonstration, synthesises a computer program based on the demonstrations, and learns behavioural models for predictive control.

The analysis of the papers presented at the AREA workshop (see Table 1) shows that, despite the many approaches proposed in the literature, it seems that there is still no consensus on the strategy to be used to specify MAS requirements. Most of the papers (5) used domain-specific languages for specifying the MAS requirements, followed by logic-based languages (2), natural languages (1), and demonstrations (1). Many research papers often do not explicitly discuss the reason that motivates the usage of a given specific language for the requirement specification. For example, many papers assume that requirements are expressed using logic-based languages (a formalism that easily supports the development of research prototypes). We believe that this is often dictated by the fact that these languages have formal semantics and are supported by verification and synthesis tools that can be reused for verifying and synthesising plans of the MAS. Others use DSLs. We believe that this choice is dictated by the need of providing solutions closer to the needs of the final users.

4.2. MAS Testing

Testing robotic and multi-agent systems is a complex activity. It starts from the unit testing level, where the units can be the single agent functionalities, to the system integration level. At the system integration level, many aspects, such as the concurrent execution of the agents, the environment integration, the control over the coordination protocol and communication management and modalities, are considered. Testing all these aspects is inherently complex, and becomes even harder when tools do not provide appropriate support.

Agent development frameworks, such as JADE [105] or Jason, support developing and testing of MAS and robotic applications. However, each of these frameworks comes with some limitations. These limitations are particularly relevant for the design and development of MAS that need to be executed in a physical distributed environment, deployed over many machines and used by humans. A concrete testing and maintenance support that covers all of these requirements is still missing. In addition, the performances of V&V techniques are not sufficient to support the requirement of distributed environments, and cannot be easily integrated in the running environment.

In this section, we summarise a set of works that considered the problem of testing MAS and robotic applications. These works have been selected based on the authors' knowledge and experience. Specifically, in this section, we considered works that are: (1) exploring current support for testing MASs, (2) analysing applications of standard testing techniques from the software engineering community, and (3) reporting how V&V approaches have been exploited in the field.

4.2.1. Support for Testing MASs

A starting point for testing an MAS is provided by the development framework itself, allowing and supporting the agent internal state inspection and the messages exchange supervision. Both JADE and Jason offer such kind of tools (the Mind Inspector in Jason, the Introspector agent, and the Sniffer in JADE), which are usually necessary to perform a manual first check of the behaviours of agents and of the overall MAS, or that can be the used to perform more automated tests and verification.

Some studies related to JADE are, for example, [106,107]. In [106], the authors proposed a solution, based on mock agents, that is presented to perform testing of a single role of an agent under successful and exceptional scenarios; in [107], the authors proposed a framework developed on JADE [105] for building and running MAS test scenarios. This framework relies on the use of aspect-oriented techniques to monitor the autonomous agents during tests and control the test input of asynchronous test cases.

In [108], the authors proposed an approach for enabling DevOps activities (that is, collaborative programming features to achieve fast and continuous deployment of complex systems) in a new framework based on JaCaMo-web IDE [109] (which is a tool related to JaCaMo (http://jacamo.sourceforge.net/ accessed: 14 May 2021), and consequently to Jason). This extension to the IDE enables for interactive facilities, such as the automatic access to the updates made to the components (agents), and the possibility to execute tests on temporary running instances (allowing the framework to check the compatibility of new changes using the real scenario, since tests are performed while the programmer types, without affecting running instances). In addition, this extension provides facilities for preventing conflicts when developers attempt to edit a resource simultaneously, and for managing versions. Thus, this is a concrete step toward offering real testing facilities to AOP.

There are some other works related to MAS testing that do not consider JADE or Jason. The SUnit framework [110] provides a model based approach based on an extension of Junit. The eCAT testing framework [111,112] supports continuous testing and automated test case generation. In [113], the authors proposed a tracing method supported by a tool implementation to capture and analyse dynamic runtime data collected by logging the behaviours of a set of agents. These solutions are ad-hoc solutions with limited practical adoption, compared to JADE or Jason. They usually rely on specific formal languages that make their use difficult, in particular in industrial applications.

4.2.2. Applications of Standard Testing Techniques from Software Engineering

Standard model-driven testing techniques coming from the SE domain (e.g., [114]) are usually difficult to be used in multi-agent applications. One of the main limitations of these techniques is that they require a model of the multi-agent application. While there are standard approaches for modelling object-oriented or service-oriented systems, standard models for MAS are less mature. Therefore, MAS design is still often performed by relying on ad-hoc solutions, which need to be standardised. Therefore, we believe that there is room for exploiting SE techniques in the Agent-Oriented Software Engineering (AOSE) area [115].

Some examples where standard SE testing approaches or techniques have been adopted for MASs platforms are reported in the following.

The BEAST methodology [116] is an example of agile testing methodology for multi-agent systems based on Behaviour Driven Development (BDD). It automatically generates test cases skeletons from BDD scenarios specifications. The BEAST framework enables testing MASs based on JADE or JADEX [117] platforms.

In [118], mutation testing is used to test Jason specifications. The authors propose a set of mutation operators for Jason, and present a mutation testing framework for individual Jason agents based on these mutation operators. In [119], the authors proposed a property-based testing (a particular form of model based testing) framework for MASs specified in Jason. Specifically, the authors proposed to replace a subset of the agents by a QuickCheck [120] state machine. This state machine interacts with the other agents by sending messages and modifying the environment, and judging whether the remaining real agents are correctly implemented by examining the messages sent to any replaced agent, and the belief perceptions that they receive.

4.2.3. Exploitation and Integration of V&V Approaches

There are many studies to check that the interaction between agents conforms to a formal specification (which is as a part of the testing activity). This is a very complex task since it involves the need to formally model and verify the protocol (as described in Section 3), and a concrete way to oversee the runtime execution, mixing together both a theoretical and an applied aspect.

A concrete example of the integration of a V&V technique directly into a development platform is proposed in [121,122], where an extension of the JADE Sniffer is used to create a monitor able to verify at runtime the MAS execution with respect to a global protocol specified using the Attribute Global Types formalism [123–125]. Since this is a JADE agent, it can be directly used in any JADE MAS, provided that the global protocol (if any) is translated into the requested language. Similarly, in [121], an extension of Jason is proposed to achieve the same monitoring.

In [39], we can find a work presenting the combination of formal verification and validation in the context of MAS verification, integrated into the MCAPL [126] framework. The verification part is obtained through the verification of the BDI agent, implemented in Gwendolen [6], against a formal model of the environment, while the corresponding validation is achieved through runtime verification, where monitors are used to verify at runtime that the real environment does not violate the assumptions made by the model checker.

4.3. Simulation Tools

Physically distributed systems are often needed in industrial and academic solutions, such as unmanned vehicles, logistic, ambient intelligent systems.

4.3.1. Simulation Tools for MASs

JADE is largely adopted in industry, due to its simplicity in modelling agents' tasks, its support for agent communication, and its extensive community support. JADE suffers from scalability issues [127,128]. In addition, it provides limited support for dynamically discovering new platforms that join the MAS at run-time. This limitation forces the usage of p2p communication, which is quite common in real world applications ([129]). However, differently from Jason, JADE was integrated within existing simulation platforms.

Before deploying the MAS, developers need to test their solutions. To be representative of a real situation, the physical environment must contain agents representing physical entities, such as vehicles, computers, and production systems. Testing these systems is usually done by first relying on simulations. We can simulate the behaviour of an MAS by relying on some stub entities, instead of using the actual MAS components. Stubs implement a logic which is similar but usually simpler than the one that will be executed by the actual agents. For example, stubs may abstract and simplify the interaction protocols used by the different agents to communicate.

Some simulators for MASs exist, and we will present them in this section, but they deal with the simulation in different ways.

In [130], the authors propose DMASF, a Python distributed simulator for large scale MASs (made of billions of agents). In this simulator, agents are implemented as specific entities using the simulator language. This means that a large number of agents can be simulated (we speak of numbers that are usually not manageable with JADE nor Jason, even using a simulation setup with many machines). However, the agents have to be re-implemented using the language of the simulator. This prevents user from testing actual code executed by agents of the MAS. Netlogo (http://ccl.northwestern.edu/netlogo/index.shtml accessed: 14 May 2021) is another simulator that can handle MAS applications with a high number of agents. However, similarly to DMASF, the logic of the MAS agents has to be re-implemented using the input language of the simulator. We do not provide a deeper analysis of these types of simulators since, in the rest of this paper, we will focus on simulators that can support JADE specifications.

The JREP platform [131] integrates JADE and Repast Simphony [132]. It solves some limitations of a similar approach presented in [133] that supports an older version of Repast and was limited by the focus on supply chain performance analysis and by an inefficient polling strategy affecting performances. The JREP platform offers an MAS development platform exploiting the JADE support for modelling the internal agent behaviour and the Repast support for simulating an environment where entities can interact in a simulation. JREP is a new development platform, where JADE agents need to be modified to implement new interfaces to be able to interact with the Repast environment. In this way, a bidirectional interface between the JADE agents and the Repast running environment is achieved, but the resulting JADE MAS is no longer able to run independently outside of the JREP platform.

In [134,135], the authors proposed the SAJaS API. This API has to be used with Repast Simphony to create, or improve, MAS based simulations enhanced with JADE-based features. Then, the "MAS Simulation to Development (MASSim2Dev)" code conversion tool transforms an MAS, developed using the SAJaS API, to a "standalone" standard JADE MAS. This tool is useful in scenarios where JADE developers need to perform tests in a simulator before distributing their JADE MAS. The architectural design of the JADE framework is based on Repast. However, since it is "JADE-like" environment, it is simpler to generate the JADE standard implementation. Unfortunately, MASSim2Dev can not manage the JADE blocking functions, and does not allow Ticker and Waker behaviours due to problems with time management.

4.3.2. Simulation Tools for Robots

Robotics systems also need to be extensively tested before deployment. Testing these systems is even more complex since robots interact with their physical environments through sensors and actuators. Therefore, to test these systems, simulators must provide reliable and accurate simulators between the robots and their environment. The variety of the environments in which the robots operate (e.g., very deep sea, disaster areas, no gravity scenarios and so on), makes the creation of accurate simulators even more complex.

In the following, we report a few examples of simulators present in the literature.

USARSim (Urban Search and Rescue Simulation) [136] is a general-purpose multi-robot simulator environment used as the simulation engine for the Virtual Robots Competition within the Robocup initiative, and has been often adopted in research activities. It presents an interface with Player [137] (a popular middleware used to control many different robots), and, thanks to this interface, the code developed within USARSim can be transparently moved to real platforms without any change (and vice versa). This simulator provides is relatively accurate: there is a close correspondence between results obtained within the simulation and the one obtained by the corresponding physical system.

MORSE (Modular Open Robots Simulation Engine) [138] is an open-source application that can be used in different contexts for the testing and verification of robotics systems. It is completely modular and can interact with any middleware used in robotics. In addition, it does not impose a format to which programmers must adapt. MORSE is designed to handle the simulation of several robots simultaneously, as a distributed application where the robotics software being evaluated can run on the same or a different computer as the simulation one. The evaluated components are executed on the target hardware and interact with the simulator with the very same protocols as the ones of the actual robots, sensors, and actuators, to make the shift from simulations to actual experiments transparent.

Gazebo [139] is a 3D dynamic multi-robot environment simulator. It is developed starting from the well known Player/Stage project [137], with the goal of enabling simulating dynamic outdoor environments and providing realistic sensor feedback, while still modelling robots as dynamic structures composed of rigid bodies connected via joints. The hardware simulated in Gazebo is designed to accurately reflect the behaviour of its physical counterpart: consequently, a client program shows an interface that is identical to the one that will be executed on the final robot. This makes Gazebo to be seamlessly inserted into the development process of a robotic system. Nowadays, it is largely adopted by the robotic community, and has a large and active supporting community.

4.4. Perspective of the Authors

Many SE approaches for MAS and robotic applications were proposed in the literature. However, many challenges still need to be addressed. In the following, we will discuss three of these challenges.

- *lack of clear guidance for the selection of the specification language to be used for the requirement specification.* The analysis of the paper presented in Section 4.1 showed the absence of a consensus on the strategy to be used to specify MAS requirements. Given the limited number of papers analysed (10), we cannot make any general claim on our observations, which should be confirmed by more extensive and in-depth studies. However, we believe that all the formalisms proposed in the literature for requirement specifications offer pro and cons, and that the research community should spend some effort in understanding when and how to use them and providing guidelines that can be used in research and practical works. We believe that our observations can pave the way for discussions and further studies on the requirement specification of multi-agent systems.
- *lack of mature testing tools for MAS and robotic applications.* The works summarised in Section 4.2 are some examples of SE techniques that support testing MAS and robotic applications. However, these techniques are supported by research prototypes that are still not mature enough to be used in industrial settings. Therefore, we believe

that more effort is needed to implement and develop mature tools that can be used in industrial applications.

- *lack of use of industrial simulators.* As discussed in Section 4.3, there are many simulators for JADE MASs. However, these simulators are still not ready for industrial usages. In addition, while there are many platforms for simulating robotics systems, the continuous innovation of available solutions from the robotic community (e.g., new sensors and actuators) is asking for more accurate simulators. It is also necessary to precisely document the usage scenarios and assumptions of each simulator, to enable developers to quickly find the best simulation platform for their needs. For this reason, we believe that research should work with integrating research solutions with real industrial products.

5. Human–Agent Interaction

Building reliable applications is of primary importance when agents and robots need to interact with humans. Robots or humans could be directly (negatively) affected by an agent's behaviour, e.g., when humans and agents are working together to achieve a goal.

To ensure reliable interactions, humans and agents need to anticipate each other's actions and reactions to some degree. They need to communicate and understand each other, as well as develop a shared understanding of their environment. In addition, humans need to trust the autonomous system.

We discuss three main approaches for realising reliable applications based on human–agent interaction, namely (i) building it right from the ground up, (ii) analysing existing interactions, and (iii) adjusting the users' expectations when necessary. In addition, the verification of human–robot interaction was discussed in Section 3.

5.1. Interaction Design

Agent interaction is usually based on interaction protocols. If humans are part of the system, the means of interaction between a user and an autonomous component have to be designed before the deployment of the application. Some approaches have been proposed to design human–robot interaction protocols.

Interaction Design Patterns (IDPs) [140] have been proposed to capture workable solutions for human–agent interaction. As design patterns are rather descriptive in nature, they allow for more flexibility in how they are actually implemented. In [141], five design patterns for eliciting self-disclosure are presented, as self-disclosure is an important part of getting acquainted, which leads to more trust and helps with long-term interactions. In their paper, children were the target audience. IDPs are also used in the framework of [142] to specify how to communicate explanations in order to improve the performance of mixed human–agent teams.

In [143], 18 guidelines for interactions between humans and AI are proposed from the perspective of the human–computer interaction field and evaluated through a user study. The guidelines include making clear what the system can do, how well it can do it, considering social norms, supporting efficient interaction, giving explanations, and adapting to the current user.

In [144], planning agents are used to coordinate in disaster response scenarios so that humans can choose to be guided by an agent. The authors also propose some guidelines for interaction design: agents should always be able to respond to the needs of the users, interactions should be rather simple with limited options, and interaction should enable transfer of control between the autonomous agent and its user. In [145], the authors proposed *adjustable autonomy*, which enables a (human) controller to switch unmanned aerial vehicles operation between manual control and autonomy, enabling them to supervise multiple entities at once and only assume control when necessary.

5.2. Modelling Mixed Human–Agent Systems

If a system can be modelled, it can be simulated (or even verified), it is easier to reason about it and reach a deeper understanding of the system behaviour. Modelling human–

agent interaction is a challenging endeavour. On the one hand, we have the autonomous nature of the agent, while, on the other hand, human behaviour tends to be unpredictable.

In [146], the authors proposed an approach to synthesise control protocols. An unmanned aerial vehicle and the operator are modelled as Markov Decision Processes (MDP), interacting via synchronised actions. MDPs allow modelling uncertainties regarding the behaviour of the human operator. The model is then augmented to a stochastic two-player game to account for non-determinism.

Another modelling approach for autonomous systems (e.g., autonomous driving) is proposed in [147]. They combine discrete event simulation with system dynamics models to simulate the effects of different system designs.

In [148], an entire robot swarm interacting with humans is considered. Challenges include which control type to use in which situation (assigning and controlling a leader, using the environment, controlling only parameters, or assigning behaviours), and also visualisation techniques to help the user understand a swarm's behaviour.

The modelling approaches for interaction design presented in this section are all created for a specific use case. More work is required to find general-purpose models that can be used in different domains.

5.3. Trust and Transparency

As systems get more and more autonomous, anticipating their behaviour becomes more challenging. Transparency can help users understand what their system is doing. In [149], the authors use a tiered transparency model based on situation awareness to show that more agent transparency leads to better trust calibration of the human operator without necessarily increasing their workload. This is further developed in [150], where the authors argue that bidirectional transparency is important. Thus, agents have to be designed to understand the plans, reasoning and future projections of their users.

In [151], the author argues that trust (in an autonomous system) at least requires a framework for recourse, the system's ability to give explanations, and verification and validation of the system.

In [150], the authors discuss further challenges regarding trust, namely how to quantify trust, how to model its evolution, how to create a logic that allows for specifications including trust, and also how to verify whether a system satisfies such a specification.

5.4. Behaviour Explanations

There are still many cases where full autonomy is not yet achievable. In these cases, humans have to take on a supervisory role. In these cases, humans may want to implicitly perform some kind of (mostly informal) verification of the agents' autonomous behaviour. One way of achieving this is giving autonomous systems the ability to explain their actions, their decisions and reasoning. While the previous paragraph handled how to enable users to get expectations which are more justified, some expectations might still be unreasonable. Behaviour explanations are one way of realigning these expectations, or even finding flaws in the autonomous logic of the system.

In [152], the author draws on the large body of work on explanations in the social sciences to infer properties of good explanations that humans will accept. They are contrastive (explaining why something happened instead of something else), selected (giving only relevant and important information) and social (considering the needs and background of the recipient).

In [153], a mechanism for answering "Why?"-questions about an agent's behaviour is implemented for the GWENDOLEN language. The questions are answered by identifying causal factors in the trace of the program, i.e., choice points, where another decision wouldn't have led to the result that needs to be explained.

A general perspective on explanations in AI is taken in [154] with a focus on systems incorporating machine learning. They give a formal definition of explainability, distinguishing it from interpretability and transparency of learning algorithms. Among others, they give possible reasons for making a system explainable (since each system does not need to be, e.g., if users have no way of reacting to an explanation), possible groups of target recipients, ways to create interpretations and the relations between these questions.

In [155], the authors take another stance on explanation. They consider explanation as a model reconciliation problem. Explaining means making the mental models of the agent and the user converge, leading to a shared understanding of their world, so that the plan of the agent appears optimal.

5.5. Perspective of the Authors

Many researchers had considered human–agent interaction problems and designed solutions for these problems. However, many challenges still need to be addressed. In the following, we will discuss three of these challenges:

- *Making human–agent interaction more reliable.* There is an increasing need for making human–agent interaction more reliable. This problem can and has to be tackled from many different research angles. Interaction design, employing foremost guidelines and design patterns have laid the foundation for reliable interaction. However, reliability is mostly targeted implicitly, which leaves a need for the incorporation of an explicit notion of reliable interaction.
- *Providing modelling formalisms that effectively enable modelling human–agent systems.* Modelling human–agent systems requires new ways of specifying formerly informal concepts, such as trust, transparency, and maybe even more exotic concepts (for a machine) such as honesty and loyalty. Of course, the ability to model such systems is closely linked to being able to verify them.
- *Making systems more understandable.* Finally, making systems more understandable, e.g., by explaining them, requires many different parts coming together. In the concrete case of improving reliability, challenges include making sure users correctly understand what they are told, systems explaining their actions truthfully and users being able to verify that, or agents being able to understand why users perceive them as unreliable and act upon that.

6. Conclusions

In this perspective paper, we have discussed the applicability of agent-based programming in robotics, presented an overview of the landscape in the verification and validation of MAS and robot systems, analysed the use of software engineering in MAS, and described the latest research in human–agent interaction. Combining knowledge coming from these research areas may lead to innovative approaches that solve complex problems related to the development of autonomous robotic systems, and there is growing interest in solutions that are at the intersection of these research areas. The AREA workshop was a successful event attended by researchers working in these areas. It was an exciting event that enabled sharing ideas, open problems, and solutions and fostering cross-disciplinary collaborations among researchers. In this work, we used the papers and discussions from this workshop to analyse some of the aspects covered in this perspective.

Our perspective provided a high-level view of current research trends. We also identified a set of challenges for each of the areas we considered. For multi-agent programming, the challenges we identified include among others, the limited set of features provided by existing agent-based languages, immature methodologies and tools, and the limited integration of agent-based technologies with other techniques. For verification and validations, the challenges include the scalability of the proposed techniques and the verification of ML components. For software engineering, the challenges include the lack of clear guidance for the selection of the specification language to be used for expressing requirements, the lack of mature testing tools for MAS and robotic applications and the lack of use of industrial simulators within research works. Finally, for human–agent interactions, the challenges include the still inadequate reliability of human–agent interaction systems, the still immature modelling support for these systems, and the lack of techniques able to make these systems more understandable. We believe that this perspective can be useful for researchers that aim at working at the intersection of these research areas.

We hope that the different research communities improve their collaboration efforts, so that the best proposals from different areas can be combined to create new and existing solutions and tools to be exploited both in academia and in industry.

Author Contributions: All authors (R.C.C., A.F., D.B., C.M. and T.A.) contributed equally to this work. All authors have read and agreed to the published version of the manuscript.

Funding: Rafael C. Cardoso and Angelo Ferrando's work in this research was supported by UK Research and Innovation, and EPSRC Hubs for "Robotics and AI in Hazardous Environments": EP/R026092 (FAIR-SPACE), EP/R026173 (ORCA), and EP/R026084 (RAIN). Claudio Menghi is supported by the European Research Council under the European Union's Horizon 2020 research and innovation programme (grant No 694277).

Conflicts of Interest: The authors declare no conflict of interest. The funders had no role in the design of the study; in the collection, analyses, or interpretation of data; in the writing of the manuscript, or in the decision to publish the results.

Abbreviations

The following abbreviations are used in this manuscript:

AI	Artificial Intelligence
AOSE	Agent-Oriented Software Engineering
BDD	Behaviour Driven Development
BDI	Belief-Desire-Intention
CAT	Capability Analysis Table
DSL	Domain-Specific Language
IDP	Interaction Design Pattern
MAS	Multi-Agent System
MDP	Markov Decision Process
PDDL	Planning Domain Definition Language
ROS	Robot Operating System
SE	Software Engineering

References

1. Bratman, M.E. *Intentions, Plans, and Practical Reason*; Harvard University Press: Cambridge, MA, USA, 1987.
2. Rao, A.S.; Georgeff, M. BDI Agents: From Theory to Practice. In Proceedings of the 1st International Conference Multi-Agent Systems (ICMAS), San Francisco, CA, USA, 12–14 June 1995; pp. 312–319.
3. Bordini, R.H.; Hübner, J.F.; Wooldridge, M. *Programming Multi-Agent Systems in AgentSpeak Using Jason (Wiley Series in Agent Technology)*; John Wiley & Sons, Inc.: Hoboken, NJ, USA, 2007.
4. Boissier, O.; Bordini, R.H.; Hübner, J.F.; Ricci, A.; Santi, A. Multi-agent Oriented Programming with JaCaMo. *Sci. Comput. Program.* **2013**, *78*, 747–761. [CrossRef]
5. Boissier, O.; Bordini, R.; Hubner, J.; Ricci, A. *Multi-Agent Oriented Programming: Programming Multi-Agent Systems Using JaCaMo*; Intelligent Robotics and Autonomous Agents Series; MIT Press: Cambridge, MA, USA, 2020.
6. Dennis, L.A.; Farwer, B. Gwendolen: A BDI Language for Verifiable Agents. In *Workshop on Logic and the Simulation of Interaction and Reasoning*; AISB: London, UK, 2008.
7. Quigley, M.; Conley, K.; Gerkey, B.; Faust, J.; Foote, T.; Leibs, J.; Wheeler, R.; Ng, A. ROS: An open-source Robot Operating System. In Proceedings of the Workshop on Open Source Software at the International Conference on Robotics and Automation, Kobe, Japan, 12–17 May 2009.
8. Wesz, R. Integrating Robot Control into the Agentspeak(L) Programming Language. Master's Thesis, Pontificia Universidade Catolica do Rio Grande do Sul, Porto Alegre, Brazil, 2015.
9. Morais, M.G.; Meneguzzi, F.R.; Bordini, R.H.; Amory, A.M. Distributed fault diagnosis for multiple mobile robots using an agent programming language. In Proceedings of the 2015 International Conference on Advanced Robotics (ICAR), Istanbul, Turkey, 27–31 July 2015; pp. 395–400. [CrossRef]
10. Onyedinma, C.; Gavigan, P.; Esfandiari, B. Toward Campus Mail Delivery Using BDI. In *Agents and Robots for Reliable Engineered Autonomy*; Electronic Proceedings in Theoretical Computer Science; Open Publishing Association: The Hague, The Netherlands, 2020; Volume 319, pp. 127–143. [CrossRef]

11. Onyedinma, C.; Gavigan, P.; Esfandiari, B. Toward Campus Mail Delivery Using BDI. *J. Sens. Actuator Netw.* **2020**, *9*, 56. [CrossRef]
12. Cardoso, R.C.; Ferrando, A.; Dennis, L.A.; Fisher, M. An Interface for Programming Verifiable Autonomous Agents in ROS. In *Multi-Agent Systems and Agreement Technologies*; Springer: Berlin/Heidelberg, Germany, 2020; pp. 191–205.
13. Crick, C.; Jay, G.; Osentoski, S.; Pitzer, B.; Jenkins, O.C. Rosbridge: ROS for Non-ROS Users. In *Robotics Research: International Symposium ISRR*; Springer: Berlin/Heidelberg, Germany, 2017; pp. 493–504.
14. Logan, B. An agent programming manifesto. *Int. J. Agent-Oriented Softw. Eng.* **2018**, *6*, 187–210. [CrossRef]
15. Bordini, R.H.; Seghrouchni, A.E.F.; Hindriks, K.V.; Logan, B.; Ricci, A. Agent programming in the cognitive era. *Auton. Agents Multi Agent Syst.* **2020**, *34*, 37. [CrossRef]
16. Cardoso, R.C.; Ferrando, A. A Review of Agent-Based Programming for Multi-Agent Systems. *Computers* **2021**, *10*, 16. [CrossRef]
17. Ziafati, P.; Dastani, M.; Meyer, J.J.; van der Torre, L. Agent Programming Languages Requirements for Programming Autonomous Robots. In *Programming Multi-Agent Systems*; Springer: Berlin/Heidelberg, Germany, 2013; pp. 35–53.
18. Luckcuck, M.; Farrell, M.; Dennis, L.A.; Dixon, C.; Fisher, M. Formal specification and verification of autonomous robotic systems: A survey. *ACM Comput. Surv. CSUR* **2019**, *52*, 1–41. [CrossRef]
19. Farrell, M.; Luckcuck, M.; Fisher, M. Robotics and integrated formal methods: Necessity meets opportunity. In *International Conference on Integrated Formal Methods*; Springer: Berlin/Heidelberg, Germany, 2018; pp. 161–171.
20. Lomuscio, A.; Qu, H.; Raimondi, F. MCMAS: An open-source model checker for the verification of multi-agent systems. *Int. J. Softw. Tools Technol. Transf.* **2017**, *19*, 9–30. [CrossRef]
21. Kouvaros, P.; Lomuscio, A. Parameterised verification for multi-agent systems. *Artif. Intell.* **2016**, *234*, 152–189. [CrossRef]
22. Čermák, P.; Lomuscio, A.; Murano, A. Verifying and synthesising multi-agent systems against one-goal strategy logic specifications. In Proceedings of the AAAI Conference on Artificial Intelligence, Austin, TX, USA, 25–30 January 2015; Volume 29.
23. Belardinelli, F.; Lomuscio, A.; Murano, A.; Rubin, S. Verification of Multi-agent Systems with Imperfect Information and Public Actions. In Proceedings of the AAMAS, São Paulo, Brazil, 5–8 May 2017; Volume 17, pp. 1268–1276.
24. Dennis, L.; Fisher, M.; Slavkovik, M.; Webster, M. Formal verification of ethical choices in autonomous systems. *Robot. Auton. Syst.* **2016**, *77*, 1–14. [CrossRef]
25. Dennis, L.A.; Fisher, M.; Webster, M.P.; Bordini, R.H. Model checking agent programming languages. *Autom. Softw. Eng.* **2012**, *19*, 5–63. [CrossRef]
26. Kashi, R.N.; D'Souza, M. *Vermillion: A Verifiable Multiagent Framework for Dependable and Adaptable Avionics*; Technical Report; IIIT-Bangalore: Bengaluru, India, 2018.
27. Meyer, B.; Baudoin, C. *Méthodes de Programmation*, 1st ed.; Eyrolles: Paris, France, 1978.
28. Karoui, O.; Khalgui, M.; Koubâa, A.; Guerfala, E.; Li, Z.; Tovar, E. Dual mode for vehicular platoon safety: Simulation and formal verification. *Inf. Sci.* **2017**, *402*, 216–232. [CrossRef]
29. Bengtsson, J.; Larsen, K.G.; Larsson, F.; Pettersson, P.; Yi, W. UPPAAL—A Tool Suite for Automatic Verification of Real-Time Systems. In *Workshop on Verification and Control of Hybrid Systems*; Springer: Berlin/Heidelberg, Germany, 1995; Volume 1066, pp. 232–243. [CrossRef]
30. Kamali, M.; Dennis, L.A.; McAree, O.; Fisher, M.; Veres, S.M. Formal verification of autonomous vehicle platooning. *Sci. Comput. Program.* **2017**, *148*, 88–106. [CrossRef]
31. Al-Saqqar, F.; Bentahar, J.; Sultan, K.; Wan, W.; Asl, E.K. Model checking temporal knowledge and commitments in multi-agent systems using reduction. *Simul. Model. Pract. Theory* **2015**, *51*, 45–68. [CrossRef]
32. Pecheur, C.; Raimondi, F. Symbolic Model Checking of Logics with Actions. In *Model Checking and Artificial Intelligence*; Springer: Berlin/Heidelberg, Germany, 2006; Volume 4428, pp. 113–128. [CrossRef]
33. Cimatti, A.; Clarke, E.M.; Giunchiglia, F.; Roveri, M. NUSMV: A New Symbolic Model Checker. *Int. J. Softw. Tools Technol. Transf.* **2000**, *2*, 410–425. [CrossRef]
34. Ancona, D.; Ferrando, A.; Mascardi, V. Parametric Runtime Verification of Multiagent Systems. In Proceedings of the AAMAS, São Paulo, Brazil, 8–12 May 2017; Volume 17, pp. 1457–1459.
35. Ferrando, A.; Ancona, D.; Mascardi, V. Decentralizing MAS Monitoring with DecAMon. In Proceedings of the Conference on Autonomous Agents and MultiAgent Systems, São Paulo, Brazil, 8–12 May 2017; ACM: New York, NY, USA, 2017; pp. 239–248.
36. Bakar, N.A.; Selamat, A. Runtime verification of multi-agent systems interaction quality. In *Asian Conference on Intelligent Information and Database Systems*; Springer: Berlin/Heidelberg, Germany, 2013; pp. 435–444.
37. Roungroongsom, C.; Pradubsuwun, D. Formal Verification of Multi-agent System Based on JADE: A Semi-runtime Approach. In *Recent Advances in Information and Communication Technology 2015*; Springer: Berlin/Heidelberg, Germany, 2015; pp. 297–306.
38. Lim, Y.J.; Hong, G.; Shin, D.; Jee, E.; Bae, D.H. A runtime verification framework for dynamically adaptive multi-agent systems. In Proceedings of the International Conference on Big Data and Smart Computing (BigComp), Hong Kong, China, 18–20 January 2016; pp. 509–512.
39. Ferrando, A.; Dennis, L.A.; Ancona, D.; Fisher, M.; Mascardi, V. Verifying and Validating Autonomous Systems: Towards an Integrated Approach. In *Runtime Verification RV*; Lecture Notes in Computer Science; Springer: Berlin/Heidelberg, Germany, 2018; Volume 11237, pp. 263–281.
40. Webster, M.; Dixon, C.; Fisher, M.; Salem, M.; Saunders, J.; Koay, K.L.; Dautenhahn, K.; Saez-Pons, J. Toward reliable autonomous robotic assistants through formal verification: A case study. *IEEE Trans. Hum.-Mach. Syst.* **2015**, *46*, 186–196. [CrossRef]

41. Sierhuis, M.; Clancey, W.J. Modeling and Simulating Work Practice: A Method for Work Systems Design. *IEEE Intell. Syst.* **2002**, *17*, 32–41. [CrossRef]
42. Holzmann, G. *Spin Model Checker, the: Primer and Reference Manual*, 1st ed.; Addison-Wesley Professional: Boston, MA, USA, 2003.
43. Holzmann, G.J. The Model Checker SPIN. *IEEE Trans. Softw. Eng.* **1997**, *23*, 279–295. [CrossRef]
44. Bérard, B.; Lafourcade, P.; Millet, L.; Potop-Butucaru, M.; Thierry-Mieg, Y.; Tixeuil, S. Formal verification of mobile robot protocols. *Distrib. Comput.* **2016**, *29*, 459–487. [CrossRef]
45. Barnat, J.; Brim, L.; Cerná, I.; Moravec, P.; Rockai, P.; Simecek, P. DiVinE—A Tool for Distributed Verification. In *Computer Aided Verification, CAV*; Springer: Berlin/Heidelberg, Germany, 2006; Volume 4144, pp. 278–281.
46. Halder, R.; Proença, J.; Macedo, N.; Santos, A. Formal verification of ROS-based robotic applications using timed-automata. In Proceedings of the IEEE/ACM FME Workshop on Formal Methods in Software Engineering (FormaliSE), Buenos Aires, Argentina, 27–27 May 2017; pp. 44–50.
47. Wang, L.; Ames, A.; Egerstedt, M. Safety barrier certificates for heterogeneous multi-robot systems. In Proceedings of the 2016 American Control Conference (ACC), Boston, MA, USA, 6–8 July 2016; pp. 5213–5218.
48. Foughali, M.; Berthomieu, B.; Dal Zilio, S.; Ingrand, F.; Mallet, A. Model checking real-time properties on the functional layer of autonomous robots. In *International Conference on Formal Engineering Methods*; Springer: Cham, Switzerland, 2016; pp. 383–399.
49. Foughali, M.; Berthomieu, B.; Dal Zilio, S.; Hladik, P.E.; Ingrand, F.; Mallet, A. Formal verification of complex robotic systems on resource-constrained platforms. In Proceedings of the IEEE/ACM International FME Workshop on Formal Methods in Software Engineering (FormaliSE), Gothenburg, Sweden, 27 May–3 June 2018; pp. 2–9.
50. Fleury, S.; Herrb, M.; Chatila, R. GenoM: A tool for the specification and the implementation of operating modules in a distributed robot architecture. In Proceedings of the IEEE/RSJ International Conference on Intelligent Robot and Systems, Innovative Robotics for Real-World Applications, Grenoble, France, 11 September 1997; pp. 842–849. [CrossRef]
51. Berthomieu, B.; Bodeveix, J.; Filali, M.; Garavel, H.; Lang, F.; Peres, F.; Saad, R.; Stöcker, J.; Vernadat, F. *The Syntax and Semantics of FIACRE*; Technical Report, Deliverable number F.3.2.11 of project TOPCASED; LAAS-CNRS: Toulouse, France, 2009.
52. Vicentini, F.; Askarpour, M.; Rossi, M.G.; Mandrioli, D. Safety assessment of collaborative robotics through automated formal verification. *IEEE Trans. Robot.* **2019**, *36*, 42–61. [CrossRef]
53. Askarpour, M.; Mandrioli, D.; Rossi, M.; Vicentini, F. SAFER-HRC: Safety analysis through formal verification in human–robot collaboration. In *International Conference on Computer Safety, Reliability, and Security*; Springer: Berlin/Heidelberg, Germany, 2016; pp. 283–295.
54. Pradella, M. A User's Guide to Zot. *arXiv* **2009**, arXiv:0912.5014.
55. Wang, R.; Wei, Y.; Song, H.; Jiang, Y.; Guan, Y.; Song, X.; Li, X. From offline towards real-time verification for robot systems. *IEEE Trans. Ind. Inform.* **2018**, *14*, 1712–1721. [CrossRef]
56. Ferrando, A.; Cardoso, R.C.; Fisher, M.; Ancona, D.; Franceschini, L.; Mascardi, V. ROSMonitoring: A Runtime Verification Framework for ROS. In *Towards Autonomous Robotic Systems*; Springer: Berlin/Heidelberg, Germany, 2020; pp. 387–399.
57. Ferrando, A.; Kootbally, Z.; Piliptchak, P.; Cardoso, R.C.; Schlenoff, C.; Fisher, M. Runtime Verification of the ARIAC Competition: Can a Robot be Agile and Safe at the Same Time? In Proceedings of the Italian Workshop on Artificial Intelligence and Robotics, Online, 25–27 November 2020; Volume 2806, pp. 7–11.
58. Sun, X.; Khedr, H.; Shoukry, Y. Formal verification of neural network controlled autonomous systems. In *ACM International Conference on Hybrid Systems: Computation and Control*; Association for Computing Machinery: Montreal, QC, Canada, 2019; pp. 147–156.
59. Askarpour, M.; Menghi, C.; Belli, G.; Bersani, M.M.; Pelliccione, P. Mind the gap: Robotic Mission Planning Meets Software Engineering. In *FormaliSE@ICSE 2020: International Conference on Formal Methods in Software Engineering*; ACM: New York, NY, USA, 2020; pp. 55–65.
60. Menghi, C.; Tsigkanos, C.; Pelliccione, P.; Ghezzi, C.; Berger, T. Specification Patterns for Robotic Missions. *IEEE Trans. Softw. Eng.* **2019**, *1*. [CrossRef]
61. Pavón, J.; Gómez-Sanz, J.J.; Fuentes, R. The INGENIAS methodology and tools. In *Agent-Oriented Methodologies*; IGI Global: Hershey, PA, USA, 2005; pp. 236–276.
62. Askarpour, M.; Rossi, M.; Tiryakiler, O. Co-Simulation of Human-Robot Collaboration: From Temporal Logic to 3D Simulation. In *Agents and Robots for Reliable Engineered Autonomy*; Electronic Proceedings in Theoretical Computer Science; Open Publishing Association: The Hague, The Netherlands, 2020; Volume 319, pp. 1–8. [CrossRef]
63. Halvari, T.; Nurminen, J.K.; Mikkonen, T. Testing the Robustness of AutoML Systems. In *Agents and Robots for Reliable Engineered Autonomy*; Electronic Proceedings in Theoretical Computer Science; Open Publishing Association: The Hague, The Netherlands, 2020; Volume 319, pp. 103–116. [CrossRef]
64. Lestingi, L.; Askarpour, M.; Bersani, M.; Rossi, M. Statistical Model Checking of Human-Robot Interaction Scenarios. In *Agents and Robots for Reliable Engineered Autonomy*; Electronic Proceedings in Theoretical Computer Science; Open Publishing Association: The Hague, The Netherlands, 2020; Volume 319, pp. 9–17. [CrossRef]
65. Edwards, V.; McGuire, L.; Redfield, S. Establishing Reliable Robot Behavior using Capability Analysis Tables. In *Agents and Robots for Reliable Engineered Autonomy*; Electronic Proceedings in Theoretical Computer Science; Open Publishing Association: The Hague, The Netherlands, 2020; Volume 319, pp. 19–35. [CrossRef]

66. Basich, C.; Svegliato, J.; Wray, K.H.; Witwicki, S.J.; Zilberstein, S. Improving Competence for Reliable Autonomy. In *Agents and Robots for Reliable Engineered Autonomy*; Electronic Proceedings in Theoretical Computer Science; Open Publishing Association: The Hague, The Netherlands, 2020; Volume 319, pp. 37–53. [CrossRef]

67. Iotti, E.; Petrosino, G.; Monica, S.; Bergenti, F. Exploratory Experiments on Programming Autonomous Robots in Jadescript. In *Agents and Robots for Reliable Engineered Autonomy*; Electronic Proceedings in Theoretical Computer Science; Open Publishing Association: The Hague, The Netherlands, 2020; Volume 319, pp. 55–67. [CrossRef]

68. Ancona, D.; Bassano, C.; Chessa, M.; Mascardi, V.; Solari, F. Engineering Reliable Interactions in the Reality-Artificiality Continuum. In *Agents and Robots for Reliable Engineered Autonomy*; Electronic Proceedings in Theoretical Computer Science; Open Publishing Association: The Hague, The Netherlands, 2020; Volume 319, pp. 69–80. [CrossRef]

69. Smith, S.C.; Ramamoorthy, S. Semi-supervised Learning From Demonstration Through Program Synthesis: An Inspection Robot Case Study. In *Agents and Robots for Reliable Engineered Autonomy*; Electronic Proceedings in Theoretical Computer Science; Open Publishing Association: The Hague, The Netherlands, 2020; Volume 319, pp. 81–101. [CrossRef]

70. Stringer, P.; Cardoso, R.C.; Huang, X.; Dennis, L.A. Adaptable and Verifiable BDI Reasoning. In *Agents and Robots for Reliable Engineered Autonomy*; Electronic Proceedings in Theoretical Computer Science; Open Publishing Association: The Hague, The Netherlands, 2020; Volume 319, pp. 117–125. [CrossRef]

71. Lami, G.; Gnesi, S.; Fabbrini, F.; Fusani, M.; Trentanni, G. *An Automatic Tool for the Analysis of Natural Language Requirements*; Technical Report; CNR Information Science and Technology Institute: Pisa, Italy, 2004.

72. Ambriola, V.; Gervasi, V. Processing natural language requirements. In Proceedings of the International Conference Automated Software Engineering, Incline Village, NV, USA, 1–5 November 1997; pp. 36–45.

73. Lazaridou, A.; Potapenko, A.; Tieleman, O. Multi-agent communication meets natural language: Synergies between functional and structural language learning. *arXiv* **2020**, arXiv:2005.07064.

74. Slhoub, K.; Carvalho, M.; Bond, W. Recommended practices for the specification of multi-agent systems requirements. In Proceedings of the IEEE Annual Ubiquitous Computing, Electronics and Mobile Communication Conference (UEMCON), New York, NY, USA, 19–21 October 2017; pp. 179–185.

75. Moreno, J.C.G.; López, L.V. Using Techniques Based on Natural Language in the Development Process of Multiagent Systems. In *International Symposium on Distributed Computing and Artificial Intelligence 2008 (DCAI 2008)*; Springer: Berlin/Heidelberg, Germany, 2009; pp. 269–273.

76. Elkholy, W.; El-Menshawy, M.; Bentahar, J.; Elqortobi, M.; Laarej, A.; Dssouli, R. Model checking intelligent avionics systems for test cases generation using multi-agent systems. *Expert Syst. Appl.* **2020**, *156*, 113458. [CrossRef]

77. Menghi, C.; Garcia, S.; Pelliccione, P.; Tumova, J. Multi-robot LTL Planning Under Uncertainty. In *Formal Methods*; Springer: Berlin/Heidelberg, Germany, 2018; pp. 399–417.

78. Lacerda, B.; Lima, P.U. Designing petri net supervisors for multi-agent systems from LTL specifications. In *International Conference on Autonomous Agents and Multiagent Systems-Volume 3*; International Foundation for Autonomous Agents and Multiagent Systems: Taipei, Taiwan, 2011; pp. 1253–1254.

79. Ghezzi, C.; Mandrioli, D.; Morzenti, A. TRIO: A logic language for executable specifications of real-time systems. *J. Syst. Softw.* **1990**, *12*, 107–123. [CrossRef]

80. Behrmann, G.; David, A.; Larsen, K.G. A tutorial on uppaal. In *Formal Methods for the Design of Real-Time Systems*; Springer: Berlin/Heidelberg, Germany; Bertinoro, Italy, 2004; pp. 200–236.

81. Menghi, C.; Tsigkanos, C.; Berger, T.; Pelliccione, P. PsALM: Specification of Dependable Robotic Missions. In Proceedings of the 2019 IEEE/ACM 41st International Conference on Software Engineering: Companion Proceedings (ICSE-Companion), Montreal, QC, Canada, 25–31 May 2019; pp. 99–102. [CrossRef]

82. Boufaied, C.; Menghi, C.; Bianculli, D.; Briand, L.; Parache, Y.I. Trace-Checking Signal-based Temporal Properties: A Model-Driven Approach. In Proceedings of the IEEE/ACM International Conference on Automated Software Engineering (ASE), Melbourne, Australia, 21–25 September 2020; pp. 1004–1015.

83. Arcaini, P.; Mirandola, R.; Riccobene, E.; Scandurra, P. MSL: A pattern language for engineering self-adaptive systems. *J. Syst. Softw.* **2020**, *164*, 110558. [CrossRef]

84. Musil, A.; Musil, J.; Weyns, D.; Bures, T.; Muccini, H.; Sharaf, M. Patterns for self-adaptation in cyber-physical systems. In *Multi-Disciplinary Engineering for Cyber-Physical Production Systems*; Springer: Berlin/Heidelberg, Germany, 2017; pp. 331–368.

85. Washizaki, H.; Uchida, H.; Khomh, F.; Guéhéneuc, Y.G. Studying software engineering patterns for designing machine learning systems. In Proceedings of the International Workshop on Empirical Software Engineering in Practice (IWESEP), Tokyo, Japan, 13–14 December 2019; pp. 49–495.

86. Washizaki, H.; Ogata, S.; Hazeyama, A.; Okubo, T.; Fernandez, E.B.; Yoshioka, N. Landscape of architecture and design patterns for iot systems. *Internet Things J.* **2020**, *7*, 10091–10101. [CrossRef]

87. Garcia, A.; Silva, V.; Chavez, C.; Lucena, C. Engineering multi-agent systems with aspects and patterns. *J. Braz. Comput. Soc.* **2002**, *8*, 57–72. [CrossRef]

88. Juziuk, J.; Weyns, D.; Holvoet, T. Design patterns for multi-agent systems: A systematic literature review. In *Agent-Oriented Software Engineering*; Springer: Berlin/Heidelberg, Germany, 2014; pp. 79–99.

89. Dastani, M.; Testerink, B. Design patterns for multi-agent programming. *Int. J. Agent-Oriented Softw. Eng.* **2016**, *5*, 167–202. [CrossRef]

90. Challenger, M.; Kardas, G.; Tekinerdogan, B. A systematic approach to evaluating domain-specific modeling language environments for multi-agent systems. *Softw. Qual. J.* **2016**, *24*, 755–795. [CrossRef]
91. Challenger, M.; Demirkol, S.; Getir, S.; Mernik, M.; Kardas, G.; Kosar, T. On the use of a domain-specific modeling language in the development of multiagent systems. *Eng. Appl. Artif. Intell.* **2014**, *28*, 111–141. [CrossRef]
92. Bauer, B.; Müller, J.P.; Odell, J. Agent UML: A formalism for specifying multiagent software systems. *Int. J. Softw. Eng. Knowl. Eng.* **2001**, *11*, 207–230. [CrossRef]
93. García, S.; Pelliccione, P.; Menghi, C.; Berger, T.; Bures, T. High-level mission specification for multiple robots. In Proceedings of the ACM SIGPLAN International Conference on Software Language Engineering, Athens, Greece, 20–22 October 2019; pp. 127–140.
94. Brazier, F.M.T.; Dunin-Keplicz, B.; Jennings, N.R.; Treur, J. Formal Specification of Multi-Agent Systems: A Real-World Case. In Proceedings of the First International Conference on Multiagent Systems, San Francisco, CA, USA, 12–14 June 1995; The MIT Press: Cambridge, MA, USA, 1995; pp. 25–32.
95. Van Langevelde, I.; Philipsen, A.; Treur, J. Formal Specification of Compositional Architectures. In *ECAI'92: European Conference on Artificial Intelligence*; John Wiley & Sons, Inc.: Hoboken, NJ, USA, 1992; pp. 272–276.
96. Bergenti, F.; Monica, S.; Petrosino, G. A scripting language for practical agent-oriented programming. In *ACM SIGPLAN International Workshop on Programming Based on Actors, Agents, and Decentralized Control*; Association for Computing Machinery: New York, NY, USA, 2018; pp. 62–71.
97. Aeronautiques, C.; Howe, A.; Knoblock, C.; McDermott, I.D.; Ram, A.; Veloso, M.; Weld, D.; SRI, D.W.; Barrett, A.; Christianson, D.; et al. *PDDL | The Planning Domain Definition Language*; Technical Report; Yale Center for Computational Vision and Control: New Haven, CT, USA, 1998.
98. Giunchiglia, F.; Mylopoulos, J.; Perini, A. The tropos software development methodology: Processes, models and diagrams. In *International Workshop on Agent-Oriented Software Engineering*; Springer: Berlin/Heidelberg, Germany, 2002; pp. 162–173.
99. Wooldridge, M.; Jennings, N.R.; Kinny, D. The Gaia methodology for agent-oriented analysis and design. *Auton. Agents Multi-Agent Syst.* **2000**, *3*, 285–312. [CrossRef]
100. Tran, N.; Beydoun, G.; Low, G. Design of a peer-to-peer information sharing MAS using MOBMAS (ontology-centric agent oriented methodology). In *Advances in Information Systems Development*; Springer: Berlin/Heidelberg, Germany, 2007; pp. 63–76.
101. Nicolescu, M.N.; Mataric, M.J. Natural methods for robot task learning: Instructive demonstrations, generalization and practice. In Proceedings of the International Joint Conference on Autonomous Agents and Multiagent Systems, Melbourne, Australia, 14–18 July 2003; pp. 241–248.
102. Verstaevel, N.; Boes, J.; Nigon, J.; d'Amico, D.; Gleizes, M.P. Lifelong machine learning with adaptive multi-agent systems. In Proceedings of the International Conference on Agents and Artificial Intelligence (ICAART 2017), Porto, Portugal, 24–26 February 2017; Volume 2, p. 275.
103. Wang, X.; Klabjan, D. Competitive multi-agent inverse reinforcement learning with sub-optimal demonstrations. In Proceedings of the 35th International Conference on Machine Learning, Stockholm, Sweden, 10–15 July 2018; pp. 5143–5151.
104. Le, H.M.; Yue, Y.; Carr, P.; Lucey, P. Coordinated multi-agent imitation learning. In Proceedings of the 34th International Conference on Machine Learning, Sydney, Australia, 6–11 August 2017; pp. 1995–2003.
105. Bellifemine, F.L.; Caire, G.; Greenwood, D. *Developing Multi-Agent Systems with JADE*; Wiley: Hoboken, NJ, USA, 2007.
106. Coelho, R.; Kulesza, U.; von Staa, A.; Lucena, C. Unit Testing in Multi-Agent Systems Using Mock Agents and Aspects. In *SELMAS'06: International Workshop on Software Engineering for Large-Scale Multi-Agent Systems*; ACM: New York, NY, USA, 2006; pp. 83–90. [CrossRef]
107. Coelho, R.; Cirilo, E.; Kulesza, U.; von Staa, A.; Rashid, A.; Lucena, C. JAT: A Test Automation Framework for Multi-Agent Systems. In Proceedings of the IEEE International Conference on Software Maintenance, Paris, France, 2–5 October 2007; pp. 425–434.
108. Amaral, C.J.; Kampik, T.; Cranefield, S. A Framework for Collaborative and Interactive Agent-Oriented Developer Operations. In *AAMAS'20: Proceedings of the International Conference on Autonomous Agents and MultiAgent Systems*; International Foundation for Autonomous Agents and Multiagent Systems: Richland, SC, USA, 2020; pp. 2092–2094.
109. Amaral, C.J.; Hübner, J.F. Jacamo-Web is on the Fly: An Interactive Multi-Agent System IDE. In *Engineering Multi-Agent Systems*; Springer: Berlin/Heidelberg, Germany, 2020; pp. 246–255.
110. Tiryaki, A.M.; Öztuna, S.; Dikenelli, O.; Erdur, R.C. SUNIT: A Unit Testing Framework for Test Driven Development of Multi-Agent Systems. In *Agent-Oriented Software Engineering VII*; Springer: Berlin/Heidelberg, Germany, 2007; pp. 156–173.
111. Nguyen, C.D.; Perini, A.; Tonella, P. Automated Continuous Testing of MultiAgent Systems. In Proceedings of the European Workshop on Multi-Agent Systems (EUMAS), Hammamet, Tunisia, 13–14 December 2007.
112. Nguyen, C.D.; Perini, A.; Tonella, P.; Miles, S.; Harman, M.; Luck, M. Evolutionary Testing of Autonomous Software Agents. In *AAMAS'09: Proceedings of the International Conference on Autonomous Agents and Multiagent Systems—Volume 1*; International Foundation for Autonomous Agents and Multiagent Systems: Richland, SC, USA, 2009; pp. 521–528.
113. Lam, D.N.; Barber, K.S. Debugging Agent Behavior in an Implemented Agent System. In *Programming Multi-Agent Systems*; Springer: Berlin/Heidelberg, Germany, 2005; pp. 104–125.

114. Zhang, Z.; Thangarajah, J.; Padgham, L. Model Based Testing for Agent Systems. In *AAMAS'09: Proceedings of the International Conference on Autonomous Agents and Multiagent Systems—Volume 2*; International Foundation for Autonomous Agents and Multiagent Systems: Richland, SC, USA, 2009; pp. 1333–1334.

115. Padmanaban, R.; Thirumaran, M.; Suganya, K.; Priya, R.V. AOSE Methodologies and Comparison of Object Oriented and Agent Oriented Software Testing. In *ICIA-16: Proceedings of the International Conference on Informatics and Analytics*; ACM: New York, NY, USA, 2016. [CrossRef]

116. Carrera, Á.; Iglesias, C.; Garijo, M. Beast methodology: An agile testing methodology for multi-agent systems based on behaviour driven development. *Inf. Syst. Front.* **2014**, *16*, 169–182. [CrossRef]

117. Braubach, L.; Pokahr, A.; Lamersdorf, W. Jadex: A BDI-Agent System Combining Middleware and Reasoning. In *Software Agent-Based Applications, Platforms and Development Kits*; Birkhäuser Basel: Basel, Switzerland, 2005; pp. 143–168.

118. Huang, Z.; Alexander, R.; Clark, J. Mutation Testing for Jason Agents. In *Engineering Multi-Agent Systems*; Springer: Berlin/Heidelberg, Germany, 2014; pp. 309–327.

119. Benac Earle, C.; Fredlund, L.Å. A Property-Based Testing Framework for Multi-Agent Systems. In *AAMAS'19: Proceedings of the International Conference on Autonomous Agents and MultiAgent Systems*; International Foundation for Autonomous Agents and Multiagent Systems: Richland, SC, USA, 2019; pp. 1823–1825.

120. Claessen, K.; Hughes, J. QuickCheck: A Lightweight Tool for Random Testing of Haskell Programs. *SIGPLAN Not.* **2000**, *35*, 268–279. [CrossRef]

121. Briola, D.; Mascardi, V.; Ancona, D. Distributed Runtime Verification of JADE and Jason Multiagent Systems with Prolog. In Proceedings of the Conference on Computational Logic, Torino, Italy, 16–18 June 2014; Volume 1195, pp. 319–323.

122. Ancona, D.; Briola, D.; Ferrando, A.; Mascardi, V. MAS-DRiVe: A Practical Approach to Decentralized Runtime Verification of Agent Interaction Protocols. In Proceedings of the Workshop "From Objects to Agents" Co-Located with 18th European Agent Systems Summer School (EASSS 2016), Catania, Italy, 29–30 June 2016; Volume 1664, pp. 35–43.

123. Mascardi, V.; Ancona, D. Attribute Global Types for Dynamic Checking of Protocols in Logic-based Multiagent Systems. *Theory Pract. Log. Program.* **2013**, *13*, 4–5.

124. Mascardi, V.; Briola, D.; Ancona, D. On the Expressiveness of Attribute Global Types: The Formalization of a Real Multiagent System Protocol. In *AI*IA 2013: Advances in Artificial Intelligence—XIIIth International Conference of the Italian Association for Artificial Intelligence*; Springer: Berlin/Heidelberg, Germany, 2013; Volume 8249, pp. 300–311.

125. Ancona, D.; Briola, D.; Ferrando, A.; Mascardi, V. Runtime verification of fail-uncontrolled and ambient intelligence systems: A uniform approach. *Intell. Artif.* **2015**, *9*, 131–148. [CrossRef]

126. Dennis, L.A. The MCAPL Framework including the Agent Infrastructure Layer and Agent Java Pathfinder. *J. Open Source Softw.* **2018**, *3*. [CrossRef]

127. Mengistu, D.; Tröger, P.; Lundberg, L.; Davidsson, P. Scalability in Distributed Multi-Agent Based Simulations: The JADE Case. In Proceedings of the Second International Conference on Future Generation Communication and Networking Symposia, Hinan, China, 13–15 December 2008; Volume 5, pp. 93–99. [CrossRef]

128. Lo Piccolo, F.; Bianchi, G.; Salsano, S. Measurement Study of the Mobile Agent JADE Platform. In Proceedings of the International Symposium on on World of Wireless, Mobile and Multimedia Networks, Buffalo-Niagara Falls, NY, USA, 26–29 June 2006; pp. 638–646.

129. Briola, D.; Micucci, D.; Mariani, L. A platform for P2P agent-based collaborative applications. *Softw. Pract. Exp.* **2019**, *49*, 549–558. [CrossRef]

130. Aprameya Rao, I.V.; Jain, M.; Karlapalem, K. Towards Simulating Billions of Agents in Thousands of Seconds. In *AAMAS'07: Proceedings of the 6th International Joint Conference on Autonomous Agents and Multiagent Systems*; ACM: New York, NY, USA, 2007.

131. Gormer, J.; Homoceanu, G.; Mumme, C.; Huhn, M.; Muller, J.P. JREP: Extending Repast Simphony for JADE Agent Behavior Components. In Proceedings of the IEEE/WIC/ACM International Conferences on Web Intelligence and Intelligent Agent Technology, Lyon, France, 22–27 August 2011; Volume 2, pp. 149–154. [CrossRef]

132. North, M.; Howe, T.; Collier, N.; Vos, J. Repast Simphony runtime system. In Proceedings of the Agent 2005 Conference on Generative Social Processes, Models, and Mechanisms, Chicago, IL, USA, 13–15 October 2005.

133. Yoo, M.J.; Glardon, R. Combining JADE and Repast for the Complex Simulation of Enterprise Value-Adding Networks. In *Agent-Oriented Software Engineering IX*; Springer: Berlin/Heidelberg, Germany, 2009; pp. 243–256.

134. Cardoso, H.L. SAjaS: Enabling JADE-Based Simulations. In *Transactions on Computational Collective Intelligence XX*; Springer: Berlin/Heidelberg, Germany, 2015; pp. 158–178. [CrossRef]

135. Lopes, J.; Cardoso, H. From simulation to development in MAS a JADE-based approach. In Proceedings of the ICAART—International Conference on Agents and Artificial Intelligence, Lisbon, Portugal, 10–12 January 2015; Volume 1, pp. 75–86.

136. Carpin, S.; Lewis, M.; Wang, J.; Balakirsky, S.; Scrapper, C. USARSim: A robot simulator for research and education. In Proceedings of the IEEE International Conference on Robotics and Automation, Rome, Italy, 10–14 April 2007; pp. 1400–1405.

137. Brian P. Gerkey, R.T.V.; Howard, A. The Player/Stage Project: Tools for Multi-Robot and Distributed Sensor Systems. In Proceedings of the International Conference on Advanced Robotics, Coimbra, Portuga, 30 June–3 July 2003; pp. 317–323.

138. Echeverria, G.; Lassabe, N.; Degroote, A.; Lemaignan, S. Modular open robots simulation engine: MORSE. In Proceedings of the IEEE International Conference on Robotics and Automation, Shanghai, China, 9–13 May 2011; pp. 46–51.

139. Koenig, N.; Howard, A. Design and Use Paradigms for Gazebo, An Open-Source Multi-Robot Simulator. In Proceedings of the IEEE/RSJ International Conference on Intelligent Robots and Systems, Sendai, Japan, 28 September–2 October 2004; pp. 2149–2154.
140. Kahn, P.H.; Freier, N.G.; Kanda, T.; Ishiguro, H.; Ruckert, J.H.; Severson, R.L.; Kane, S.K. Design patterns for sociality in human–robot interaction. In Proceedings of the ACM/IEEE International Conference on Human Robot Interaction, Amsterdam, The Netherlands, 12–15 March 2008; pp. 97–104.
141. Ligthart, M.; Fernhout, T.; Neerincx, M.A.; van Bindsbergen, K.L.; Grootenhuis, M.A.; Hindriks, K.V. A child and a robot getting acquainted-interaction design for eliciting self-disclosure. In Proceedings of the International Conference on Autonomous Agents and Multiagent Systems, Montreal, QC, Canada, 13–17 May 2019; pp. 61–70.
142. Neerincx, M.A.; van der Waa, J.; Kaptein, F.; van Diggelen, J. Using perceptual and cognitive explanations for enhanced human–agent team performance. In *International Conference on Engineering Psychology and Cognitive Ergonomics*; Springer:Berlin/Heidelberg, Germany, 2018; pp. 204–214.
143. Amershi, S.; Weld, D.; Vorvoreanu, M.; Fourney, A.; Nushi, B.; Collisson, P.; Suh, J.; Iqbal, S.; Bennett, P.N.; Inkpen, K.; et al. Guidelines for human-AI interaction. In Proceedings of the Chi Conference on Human Factors in Computing Systems, Glasgow, UK, 4–9 May 2019; pp. 1–13.
144. Ramchurn, S.D.; Wu, F.; Jiang, W.; Fischer, J.E.; Reece, S.; Roberts, S.; Rodden, T.; Greenhalgh, C.; Jennings, N.R. Human–agent collaboration for disaster response. *Auton. Agents Multi-Agent Syst.* **2016**, *30*, 82–111. [CrossRef]
145. Orsag, M.; Haus, T.; Tolić, D.; Ivanovic, A.; Car, M.; Palunko, I.; Bogdan, S. Human-in-the-loop control of multi-agent aerial systems. In Proceedings of the 2016 European Control Conference (ECC), Aalborg, Denmark, 29 June–1 July 2016; pp. 2139–2145.
146. Feng, L.; Wiltsche, C.; Humphrey, L.; Topcu, U. Synthesis of human-in-the-loop control protocols for autonomous systems. *IEEE Trans. Autom. Sci. Eng.* **2016**, *13*, 450–462. [CrossRef]
147. Cummings, M.; Clare, A. Holistic modelling for human-autonomous system interaction. *Theor. Issues Ergon. Sci.* **2015**, *16*, 214–231. [CrossRef]
148. Kolling, A.; Walker, P.; Chakraborty, N.; Sycara, K.; Lewis, M. Human interaction with robot swarms: A survey. *IEEE Trans. Hum.-Mach. Syst.* **2015**, *46*, 9–26. [CrossRef]
149. Selkowitz, A.; Lakhmani, S.; Chen, J.Y.; Boyce, M. The effects of agent transparency on human interaction with an autonomous robotic agent. In *Human Factors and Ergonomics Society Annual Meeting*; SAGE Publications Sage CA: Los Angeles, CA, USA, 2015; Volume 59, pp. 806–810.
150. Schaefer, K.E.; Straub, E.R.; Chen, J.Y.; Putney, J.; Evans, A.W., III. Communicating intent to develop shared situation awareness and engender trust in human–agent teams. *Cogn. Syst. Res.* **2017**, *46*, 26–39. [CrossRef]
151. Winikoff, M. Towards trusting autonomous systems. In *International Workshop on Engineering Multi-Agent Systems*; Springer: Berlin/Heidelberg, Germany, 2017; pp. 3–20.
152. Miller, T. Explanation in artificial intelligence: Insights from the social sciences. *Artif. Intell.* **2019**, *267*, 1–38. [CrossRef]
153. Koeman, V.J.; Dennis, L.A.; Webster, M.; Fisher, M.; Hindriks, K. The "Why did you do that?" Button: Answering Why-questions for end users of Robotic Systems. In *International Workshop on Engineering Multi-Agent Systems*; Springer: Berlin/Heidelberg, Germany, 2019; pp. 152–172.
154. Rosenfeld, A.; Richardson, A. Explainability in human–agent systems. *Auton. Agents Multi-Agent Syst.* **2019**, *33*, 673–705. [CrossRef]
155. Chakraborti, T.; Sreedharan, S.; Zhang, Y.; Kambhampati, S. Plan explanations as model reconciliation: Moving beyond explanation as soliloquy. In Proceedings of the International Joint Conference on Artificial Intelligence, Melbourne, Australia, 19–25 August 2017; pp. 156–163.

Journal of *Sensor and Actuator Networks*

Article

Toward Campus Mail Delivery Using BDI [†]

Chidiebere Onyedinma [1], Patrick Gavigan [2,*] and Babak Esfandiari [2,*]

[1] School of Electrical Engineering and Computer Science, University of Ottawa, Ottawa, ON K1N 6N5, Canada; conye066@uottawa.ca
[2] Department of Systems and Computer Engineering, Carleton University, Ottawa, ON K1S 5B6, Canada
* Correspondence: patrickgavigan@sce.carleton.ca (P.G.); babak@sce.carleton.ca (B.E.)
† This paper is an extended version of our paper published in In Cardoso, R.C.; Ferrando, A.; Briola, D.; Menghi, C.; Ahlbrecht, T. In Proceedings of the First Workshop on Agents and Robots for reliable Engineered Autonomy (AREA 2020), Virtual Event, 4 September 2020; pp. 127–143.

Received: 30 October 2020; Accepted: 29 November 2020; Published: 8 December 2020

Abstract: Autonomous systems developed with the Belief-Desire-Intention (BDI) architecture tend to be mostly implemented in simulated environments. In this project we sought to build a BDI agent for use in the real world for campus mail delivery in the tunnel system at Carleton University. Ideally, the robot should receive a delivery order via a mobile application, pick up the mail at a station, navigate the tunnels to the destination station, and notify the recipient. In this paper, we discuss how we linked the Robot Operating System (ROS) with a BDI reasoning system to achieve a subset of the required use casesand demonstrated the system performance in an analogue environment. ROS handles the connections to the low-level sensors and actuators, while the BDI reasoning system handles the high-level reasoning and decision making. Sensory data is sent to the reasoning system as perceptions using ROS. These perceptions are then deliberated upon, and an action string is sent back to ROS for interpretation and driving of the necessary actuator for the action to be performed. In this paper we present our current implementation, which closes the loop on the hardware-software integration and implements a subset of the use cases required for the full system. We demonstrated the performance of the system in an analogue environment.

Keywords: belief-desire-intention (BDI); jason; robot operating system (ROS); robotic agents

---

1. Introduction

An autonomous agent can be defined as a system that pursues its own agenda, affecting what it senses in the future, by sensing the environment and acting on it over time [1]. Autonomous systems should be designed in such a manner that they can intelligently react to ever-changing environments and operational conditions. Given such flexibility, they can accept goals and set a path to achieve these goals in a self-responsible manner while displaying some form of intelligence.

The Belief-Desire-Intention (BDI) framework is meant for developing autonomous agents, in that it defines how to select, execute and monitor the execution of user-defined plans (Intentions) in the context of current perceptions and internal knowledge of the agent (Beliefs) in order to satisfy the long-term goals of the agent (Desires). However, so far, very few applications of BDI have been observed outside of simulated or virtual environments. In this paper, we describe how we built our autonomous robot that uses BDI (and specifically, the Jason implementation of the BDI AgentSpeak language) and Robot Operating System (ROS) to eventually deliver interoffice mail in the Carleton University campus tunnels. The Carleton tunnel system allows people to go from one campus building to another without having to face Ottawa's harsh winters and makes for a more controlled environment for our robot to navigate. However, being underground also means that access to Global Navigation Satellite System (GNSS) signals, such as Global Positioning System (GPS), is not possible, and internet access is limited to certain areas. In this context, ultimately the robot will have to know where it is and where to deliver mail, but there are also some sub-goals, like obstacle avoidance and battery recharge, which it might have to achieve in order to get to its main goal.

Our use of BDI for this work is two fold. First, BDI provides a good goal-oriented agent architecture that is resilient to plan failure and changes to context. It also supports the notion of shorter-term and longer-term plans that can be organized so as not to conflict with each other. Granted, BDI may not necessarily be the perfect ad hoc solution for agent-based robotics, but there is just not enough literature to demonstrate the appropriateness of BDI (or lack thereof) in robotics.

There are alternative agent architectures to BDI that are available, for example the subsumption architecture [2,3]. Although it is possible that the robotic behaviours implemented in this paper could have implemented the same robot using subsumption, our longer-term goals for this project would likely make the use of other architectures more difficult. BDI is goal-directed whereas in the case of the subsumption architecture, the agent behaviour emerges from the various layers built into the agent [4].

In the remainder of the paper, we first provide some background on BDI and ROS (Section 2), and related work on known implementations of agent-based robots (Section 3). We then describe our overall hardware and software architecture (Section 4). Next, we describe in more detail the hardware and software implementation (Section 5) followed by an evaluation of the architecture (Section 6) and a discussion of the lessons learned using our method for agent-based robotics (Section 7). Our conclusion (Section 8) provides a summary of the key accomplishments presented in this paper as well as our plans for future work. Additional details with respect to the hardware implementation, specifically related to our connections between our computer and the robot's power system and our line sensor circuit, used for path following, are provided (Appendix A).

2. Background

We provide background on BDI and the AgentSpeak language in Section 2.1. We then introduce ROS in Section 2.2.

2.1. Belief-Desire-Intention Architecture

The principles that underpin BDI originated in the 1980s cognitive science theory as a means of modelling agency in humans [5]. Since that time, this model has been applied in the development of software agents as well as the field of Multi Agent System (MAS). An example of a popular implementation of applying BDI to agent reasoning is Jason [1,6]. In Jason, an agent's initial belief base, goals and plans are specified using AgentSpeak.

In BDI systems, a software agent performs reasoning based upon internally held beliefs, stored in a belief base, about itself and the task environment. The agent also has objectives, or desires, that are provided to it, as well as a plan base, which contains various means for achieving goals depending on the agent's context. The agent's reasoning cycle consists of first perceiving the task environment and receiving any messages. From this information, the agent can then decide on a course of action suitable to the context provided by those perceptions, the agent's own beliefs, messages received, and desires. Once this course of action has been selected, we can say that the agent has set an intention for itself. These plans can include updating the belief base, sending messages to other agents, and taking some action. As the agent continues to repeat its reasoning cycle, it can reassess the applicability of its intentions as it perceives the environment, dropping intentions that are no longer applicable [1,6].

Agents developed for BDI systems using Jason are programmed using a language called AgentSpeak. This is a logic-based programming language that bears similarities to Prolog. The syntax provides a means for specifying initial beliefs for the agent to have, rules that can be applied for reasoning as well as plans that can be executed. The Extended Backus–Naur form (EBNF) description of AgentSpeak can be found in Appendix A.1 of [1]. Here, we provide a brief overview of AgentSpeak.

In general, AgentSpeak plans have the form of:

```
triggeringEvent : context <- body.
```

A triggering event is the addition or deletion of a belief, achievement goal, or a test goal. The syntax of these triggers takes the form of predicates. To differentiate goals from beliefs, achievement goals begin with an exclamation mark (!) and test goals begin with a question mark (?). Triggers that are based on the addition or deletion of a belief or goal begin with a positive (+) or negative (−) sign respectively. An achievement goal is used for providing the agent with an objective with respect to the state of the environment whereas a test goal is generally used for querying the state of the environment. The context is a set of conditions that must be satisfied for the plan to be applicable based on the state of the agent's belief base. This is a logical sentence that can use both beliefs as well as previously defined rules. The body includes the instructions for the agent to follow for executing the plan. The plan body can include the addition or deletion of beliefs and/or goals as well as actions for the agent to perform [1].

Listing 1 provides an example of an AgentSpeak plan. Here, we have a simple plan for the achievement goal of `!goTo(LOCATION)`. The variable `LOCATION`, interpreted as a variable due to the capitalization of the first letter, specifies the agent's destination. This specific plan is meant for the context where the agent has arrived at the destination. Therefore, the plan context requires the agent have `direction(LOCATION,arrived)` be a logical consequence of its belief base. This can either be the result of this predicate being perceived, the belief being communicated to the agent, or adopted through plan execution. This context is applicable when `LOCATION` unifies the goal with the agent's belief. For example, during execution, the agent could adopt the goal of `!goTo(post1)`. In this case, `post1` specifies the specific location that the agent needs to travel to. This plan would become applicable if the predicate `direction(post1,arrived)` is a logical consequence of the belief base. The body of the plan is for the agent to execute the stop action, specified using the `drive(stop)` predicate.

Listing 1. Example AgentSpeak program.

```
+!goTo(LOCATION)
    :    direction(LOCATION,arrived)
    <-   drive(stop).
```

BDI enables agent-based systems to perform reasoning based upon beliefs in order to enable the agent to achieve its goals, making it an attractive way for the implementation of autonomous systems. However, for the application in this paper, this language needs a means to communicate to its sensors and actuators. ROS, described in the next section, provides the middleware necessary to connect Jason to hardware.

2.2. Robot Operating System (ROS)

ROS is a package for developing software for robotic applications [7]. ROS has an active community supporting a variety of robotic platforms, sensors, and actuators. By building robotic applications that are compatible with ROS, developers enable their applications to be compatible with other devices and software nodes supported by the community. This allows developers to focus on the implementation of individual nodes and enables flexibility to use one of many available nodes that are compatible with ROS. For example, various hardware component developers have made ROS nodes available, allowing systems developers to use those modules without concern as to how those nodes are implemented in detail.

ROS operates using a tuple-space architecture. Various software nodes publish and subscribe to various topics using socket-based communications instead of communicating with other nodes directly. This removes the need for developers of individual nodes to concern themselves with which nodes they are interacting with, they need only concern themselves with the topics that they use. This is managed using a central master node which has the role of brokering peer-to-peer connections between nodes that publish and subscribe to the same topics. ROS also provides functionality for recording run-time data, which can be used for diagnostics.

3. Related Work

Although there are many examples of research on software agents and the use of BDI, this review of related work focuses on the application of BDI to robotics where the development was targeted toward real-world applications. We will also discuss work that sought to use BDI agents with ROS. This work is discussed in Section 3.1. Next, in Section 3.2, we discuss how our approach in the context of the related work.

3.1. BDI for Robotic Applcations

The Australian military conducted research into the use of BDI for controlling a fixed-wing Unmanned Aerial Vehicle (UAV) called a Codarra Avatar. As part of this project, they developed both the "Automated Wingman", a graphical programming environment where pilots could provide mission-specific programming for a UAV, as well as a BDI-based flight controller for the UAV itself. The intent of this research was to enable pilots, who may not have programming skills, to provide mission parameters in a way more natural to them using the military's Observe Orient Decide Act (OODA) loop. The authors proposed that the OODA loop could be approximated using BDI. Successful flight tests were performed using these systems in the mid 2000s, although it is unclear if any follow-on research was conducted [8,9].

A more recent example of BDI being used for controlling a drone was provided by Menegol [10,11]. Their implementation used the JaCaMo framework [12], which includes Jason. A video of their UAV flying is available online [13]. This work is currently being extended to use the ROS as the core of the architecture [14,15]. Their approach is to build a linkage between ROS and Jason, where Jason agents can run actions by passing messages to modules in ROS and receive perceptions by receiving messages from other modules. The perceptions and actions are defined using manifest files that specify the properties and parameters of the messages. A more generalized version of this project called jason_ros, for other types of integration between Jason and ROS for other applications has evolved out of the work with UAVs [16]. This is similar to other efforts to link ROS to Jason, such as rason [17], and JROS [18], although it is unclear if these efforts are related to this project.

Taking another approach using Python, the Python RObotic Framework for dEsigning sTrAtegies (PROFETA) library implements BDI and the AgentSpeak language for use with autonomous robots [19]. They are interested in determining if Agent Oriented Programming (AOP) can be implemented with Python for simpler robotic implementations. In their paper, the authors used the Eurobot challenge as well as a simulated warehouse logistics robot scenario as case studies. In the Eurobot challenge, the robot must sort objects in the environment while also working in the presence of other, uncooperative, robots [20].

The ARGO project [21] has interfaced Jason agents with an Arduino using a library called Javino [22]. Javino is a Java library for controlling Arduino computers from Java programs that was specifically designed with the intention of using it to control a robot using Jason programs. The authors of the ARGO paper claim to not be tied to specific hardware or a specific AOP language, such as AgentSpeak [21,22].

Alzetta and Giorgini contributed work toward a real-time BDI system connected to ROS 2 [23,24]. Their implementation uses a custom built BDI engine, implemented in C++, which supports soft real-time constraints. The agent's desires are encoded with soft real-time deadlines for when they need to be achieved. The plans in the agent's plan library include the execution time for that plan. The agent reasoning system can then reason about the priority of desires, time constraints and execution time when performing plan selection.

Dennis et al. explored the use of rational agents implemented with GWENDOLEN and several robotic applications [25]. A key feature of their implementation was the use of an "abstraction engine" for performing the translation between the agent, the "physical engine" and the "continuous engine", which were responsible for the interface with the real world (or simulated) sensors and actuators of the robot. They used this method to address the challenge of using an agent reasoning system, which operates using "discrete first order logic predicates", to control a robot in the real-world which can include continuous sensor signals. Their concern was that such continuous signals could overwhelm the BDI reasoner.

Cardoso et al. interfaced BDI agents, implemented in GWENDOLEN, with ROS as described in [26]. Their implementation uses the rosbridge protocol to connect their agent reasoner with ROS. They also experimented with connecting Jason to ROS using the protocol approach. Their choice of GWENDOLEN was motivated by their desire to make use of the Agent Java PathFinder (AJPF) model checking tool. They also highlight two interesting issues related to linking agents with robots. First, the concern that the sensors may overwhelm the reasoner, as the sensors may generate data faster than the agent can handle it. They address this issue by proposing the use of filters to the sensor data to prevent the agent from being overwhelmed. They also mention an issue with implementing actions using synchronous service routines in ROS, which would cause the agent to wait for the action to be completed before continuing the reasoning cycle. Their proposed approach is to use an external handler for executing longer term actions. This handler provides updates to the agent, which can in turn command the handler to stop or continue the longer-term action, as necessary.

The authors' own related work includes the Simulated Autonomous Vehicle Infrastructure (SAVI) project, which aimed to develop an architecture for simulating autonomous agents implemented using a Jason BDI [27,28]. Among its key features is the decoupling of the agent reasoning cycle from the simulation time cycle, enabling the simulated agents to run in their own time. The agent's perceptions and actions passed between the simulated agent body running in a separate thread and decoupled from the agent's reasoning cycle. Although this system was targeted toward a simulated environment, the design was intended to be useful for application to robotic agents, not only simulated agents.

3.2. Comparison to Related Work

Our goal is to use an established BDI system, namely Jason, in an ecosystem for various robotic platforms (ROS) and enable the use of agent systems to solve real-world problems using robotics, taking advantage of ROS' ecosystem of publishers and subscribers. As mentioned in Section 3.1, while there are some projects that have sought to control real-world robotics using BDI reasoning systems, there are a limited number of works in this area. Here, we will discuss the difference in our approach to those discussed in the related work.

In the case of the Codarra avatar agent, although it is very interesting, it does not seem to be openly available. Other work, such as PROFETA, uses a Python based BDI, as opposed to the more commonly used Jason. Our work is more similar in motivation to the efforts to link Jason and ROS, although our implementation of the connection between ROS and Jason is quite different. In our case, the BDI reasoning system is built as a stand-alone program with rosjava using Jason as a library, without the use of an external middleware.

Our approach does have similarities with the approach taken by Dennis et al. with respect to the use of abstraction engines. The perception and action translators that we use could also be thought of as abstraction engines. A key difference in our approach is that we do not have a single abstraction engine for the agent to interact with but several translators for various sensors and actuators, although they could also be implemented as a single node. The idea here is that a developer could add or remove such translators (as well as the underlying sensors and actuators) as necessary, without necessarily needing to rework the unaffected nodes. This allows for more flexibility in reconfiguring the system to support new sensors and actuators. The challenge, however, is that we need to handle issues with asynchronous sensor data becoming asynchronous perceptions. This challenge was mentioned by Dennis et al. as well as by Cardoso et al. They also highlighted the challenges of potentially overwhelming the reasoner with frequent sensor updates and the issue of the agent waiting synchronously while actions are completed, possibly stalling the reasoning cycle while an action that takes a long time is executed. As mentioned in the related work, they proposed an external handler for executing these longer-term actions. This handler provides updates to the agent, which can in turn command the handler to stop or continue the longer-term action, as necessary. This bears some similarity with the our approach, which uses the action translator in this service handler capacity, however our reasoning system does not wait for the action to be completed, only for the action to be passed to the state synchronization module within SAVI ROS BDI. The trade-off of this design difference is discussed in greater detail in Section 7.

4. Architecture

This section outlines the architecture of the mail delivery robot. The robot is intended to function on an on-demand basis, where a mail-sending user would summon the robot to collect mail, like how users request rides using ride-sharing apps. The robot would then autonomously navigate to a nearby mail collection and delivery location to collect the item from the user. Once the mail has been collected, the robot would then autonomously navigate to the mail delivery location and alert the receiver that there is mail for them to receive. The receiver would then meet the robot at another mail collection and delivery location. For the purposes of this early stage prototype, the mail delivery locations, and any other points of interest are indicated using a Quick Response (QR) code, and the robot paths are marked using a line for the robot to follow. Removing the need for instrumenting the environment will be discussed in the future work, in Section 8.2.

First, in Section 4.1, we examine the task environment that the robot will operate in. We then discuss the hardware configuration in Section 4.2. The software architecture is discussed in Section 4.3.

4.1. Environment

The eventual task environment for the robot is the tunnel system that connects the buildings of Carleton University. This provides our robot with an indoor space which connects to almost every building on campus with no weather to deal with and smooth floors to drive on. Although these are attractive features of the tunnel system, there are some drawbacks. First, the tunnels do not have consistent wireless internet coverage, although there are locations where there is reliable network access. The tunnels also have lower lighting levels than typical office environments, providing a potential challenge to the design. Finally, in the tunnel there is no access to GNSS signals, such as GPS, meaning that the robot will need to determine its location another way. At our current stage of development, we have focused our development and testing efforts in an analogue environment where we have focused our testing on the performance of the agent reasoning system in preparation for our planned work with the actual tunnels.

4.2. Hardware

The hardware configuration of the mail delivery robot is shown in Figure 1. The mail delivery robot is primarily implemented using an iRobot Create2, which is the development version of the Roomba vacuum cleaning robot, without the vacuum-cleaning components. This robot can be controlled using a command protocol over a serial interface [29] and can also be used to provide power to other connected devices. A Raspberry Pi 4 computer was attached to the robot and connected via a serial cable and powered from the robot's battery using a power adapter. Furthermore, connected via a serial connection are a camera and a line sensor used for detecting a line on the floor of the tunnels.

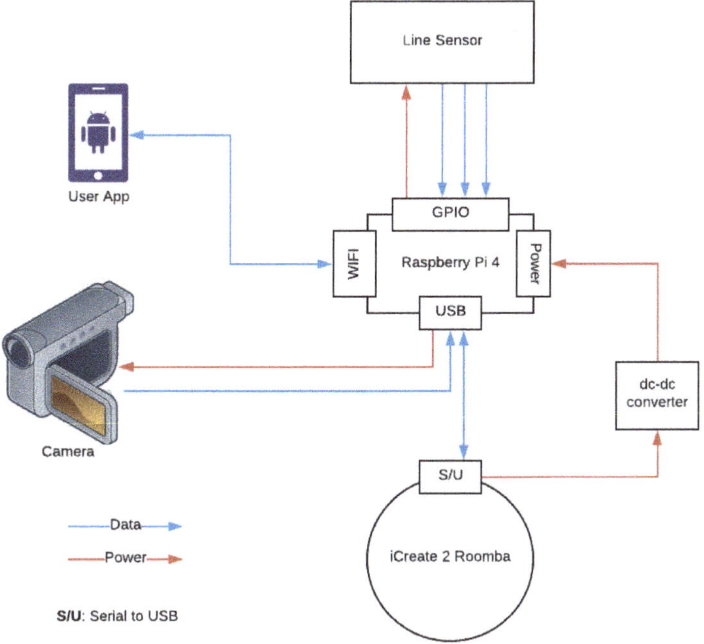

Figure 1. Mail delivery robot hardware.

4.3. Software

The control software is implemented using a set of modules connected via ROS, as shown in Figure 2. The reasoning system for this robot, inspired by the SAVI project [27,28], decouples the reasoning cycle from the interface to the sensors and actuators using a state synchronization module. The internal reasoning system for this project, called SAVI ROS BDI, and shown in Figure 3, is inspired by the original SAVI configuration. Implemented in Java, using the rosJava package [30] and the Jason BDI engine, this module connects to ROS directly, subscribing to perceptions and inbox messages and publishing actions and outbox messages as required. Again, the state synchronization module is important as perceptions and messages can arrive at any time, decoupled from the reasoning cycle of the agent. This is set up in three main components: The ROS connectors, the state synchronization module, and the agent core. The ROS connectors are responsible for subscribing to either perceptions or inbox messages, or publishing actions or outbox messages, each in their own thread of execution. These are connected to the state synchronization module, which manages queues or messages in and out of the agent as well as perceptions and actions in and out of the agent. The agent core, which runs the agent reasoning cycle in a separate thread of execution, checks for perceptions and inbox messages at the beginning of the reasoning cycle. Then, the agent decides on an appropriate course of action and then updates the agent state with new outbox messages and actions which need to be executed. The agent behaviour is defined by an AgentSpeak file which is parsed by the reasoning system at start-up, making this module fully platform agnostic: there are no assumptions about the underlying hardware, capabilities, or mission of the agent in the implementation of this system. This agent reasoning system is available at [31].

The Create2 robot platform can use the `create_autonomy` package available in ROS, which connects to an underlying C++ library called `libcreate` to ROS, publishing the data from various sensors as ROS topics and subscribing to topics related to the various commands available to the robot [32]. Furthermore, connected in this way are drivers for the QR camera and photodiode line sensor, which each publish their data as ROS topics. SAVI ROS BDI is similarly connected to ROS. Lastly, as required by SAVI ROS BDI, are the application node translators, which translate sensor data into AgentSpeak perceptions and conversely translate action commands in AgentSpeak to the relevant topics being subscribed to by the `create_autonomy` package. A user interface, which publishes to the inbox and subscribes to the outbox is included for the user to be able to communicate with the agent using Jason's agent communication mechanism. Lastly, an AgentSpeak program is provided to the reasoning system, which defines the behaviour of the agent. The implementation of the perception and action translators, the drivers for the QR camera and the line sensor, and the AgentSpeak program are discussed in Section 5. The implementation of these programs is available at [33].

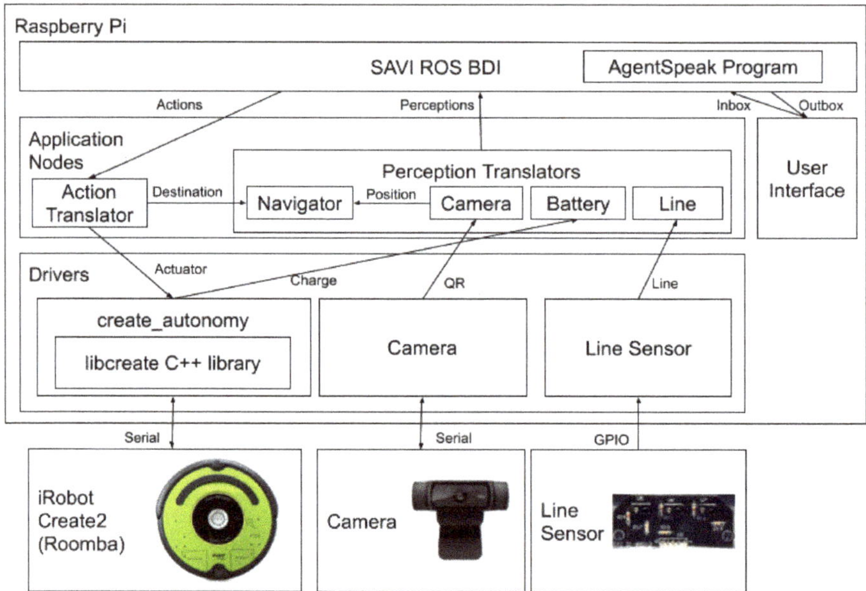

Figure 2. Mail delivery robot software architecture (robot image credit [29], camera image credit: [34]).

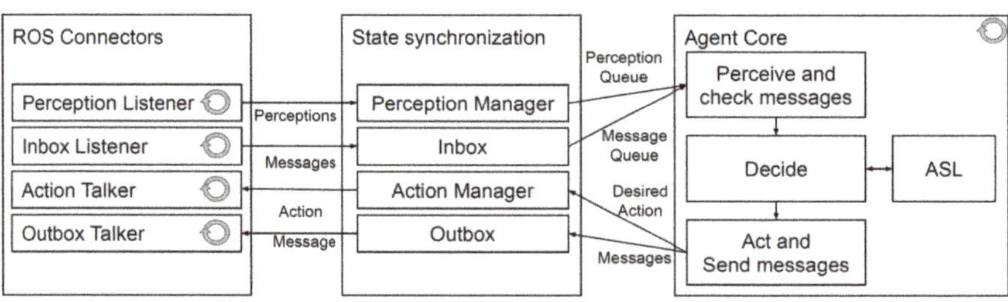

Figure 3. SAVI ROS BDI internal architecture.

5. Implementation

This section discusses the implementation of the various aspects of the system, shown in Figure 4. The source code for this project can be found on GitHub [33,35]. First, we discuss how the agent perceives the battery's state of charge in Section 5.1. Next, we discuss the means of maneuvering the robot using line sensing in Section 5.2. As the robot is expected to operate in an environment without access to GNSS signals, the robot uses a system based on QR floor markers for determining its position.

This is discussed in Section 5.3. The user interface is discussed in Section 5.5. The action translator, which handles the implementation of the robot's actuators is explained in Section 5.6. The details of the implementation of the agent behaviour, in AgentSpeak are provided in Section 5.7.

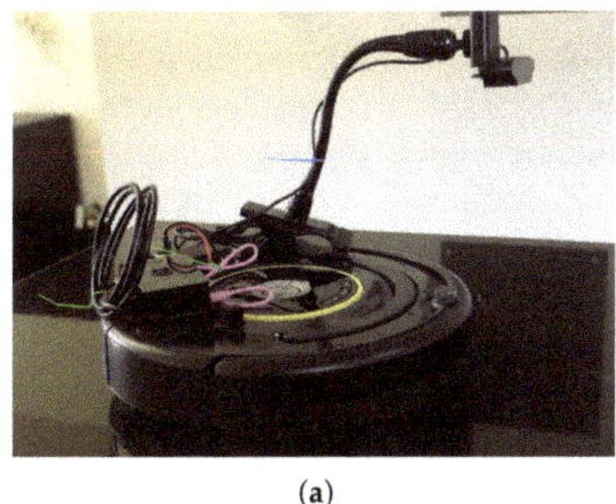

(a)

(b)

Figure 4. Assembled robot prototype (side view: (**a**), top view: (**b**)).

5.1. System Power

The robot and control computer (a Raspberry Pi 4) were both powered using the iRobot Create2's power system. The method of connecting these components is described in Appendix A.1. With the robot and computer successfully powered by the robot's power supply, it is necessary for the reasoning system to have awareness of the battery charge state so that it can report to a charging station if necessary. The `create_autonomy` package regularly publishes a ROS topic called `battery/chargeratio` which indicates the percentage of charge left on the battery based on its capacity. The perception translator node, implemented in Python, subscribes to this topic and publishes a `battery(full)`, `battery(ok)`, or `battery(low)` string to the `perceptions` ROS topic. If the battery has greater than 99 % charge remaining, the `battery(full)` and `battery(ok)` perceptions are published. If the battery has less than 25 % charge, the `battery(low)` perceptions is published. If the battery has between 25 % and 99 % charge, the `battery(ok)` perception is issued. These perceptions enable the robot to drop its intentions and seek a charging station when needed as well as resume its mission when charging is complete. These robot behaviours are explained in more detail in Section 5.7.

5.2. Maneuvering with Line Sensing

As our robot operates in an indoor environment without the support of GNSS systems for navigation, a simple means of moving through the tunnels and navigation was required. As an initial implementation, a line sensor was used for the robot to follow lines on the tunnel floor. The implementation of this line sensor is discussed in Appendix A.2.

With the line sensor hardware implemented, we needed to consider how the signals would be sent to the BDI reasoning system. A ROS node was implemented for measuring the line sensor signal and publishing it for the reasoner. This node ran in a 10 Hz loop, implemented using ROS's `rospy.Rate()` and `rospy.sleep()` functions, and interfaces with the hardware via the Raspberry Pi's General-Purpose Input/Output (GPIO) library. The software monitors if the signals from the GPIO pins are HIGH or LOW, indicating if the diodes of the line sensor are detecting the line under them. The line sensor driver interprets the signals from the sensor to estimate if the line was centered under the sensor, to the left or right side of the sensor, lost, or visible across the whole sensor. This information was published to the `perceptions` topic. The content of these messages was formatted as logical predicates which are useful for the reasoning system. These include `line(center)`, `line(left)`, `line(right)`, `line(across)`, and `line(lost)`. These perceptions were then received by the BDI reasoner and interpreted as part of the agent reasoning cycle, discussed in Section 5.7. The line sensor software node was implemented together with the QR node in order to ensure that the perceptions for the line were generated together with any position data. Location sensing with QR codes is discussed in Section 5.3.

5.3. Location Sensing with QR Codes

As the tunnel system in which the robot is expected to operate has no access to external navigation systems, such as GNSS, it was necessary for the robot to have another means of identifying its location. This was accomplished by posting QR codes along the path of the robot but without obstructing the line track that the robot would be following. The camera used for scanning the codes was also positioned on the left side of the robot and ten inches from the floor because of its focal length; this was to enable the camera to capture the code properly.

The QR code is scanned using software responsible for managing the camera. Implemented in Python, the camera driver scans for QR codes at a rate of 10 Hz. When a code is detected, the location code included in the image is logged. A perception is prepared and published to the `perceptions` topic as well as to the `postPoint` topic. The format of this perception is: `postPoint(CURRENT, PREVIOUS)` where `CURRENT` is the current scanned location code, and `PREVIOUS` is the previously scanned location code. This predicate is received by the BDI reasoning system and processed using the AgentSpeak rules discussed in more detail in Section 5.7. It is also received by the navigation module, discussed in Section 5.4. As mentioned in Section 5.2, this node was implemented together with the line sensor node to ensure that the perceptions associated with the line sensor and location sensor were perceived together by the reasoning system.

5.4. Navigation Module

The navigation module uses an A* search to find the best path to the destination from the current location. Implemented in Python, this module subscribes to the location sensing module, reading the `postPoint` data. This module also subscribes to `setDestination`, monitoring for a command specifying the agent's desired destination. The map of the environment, shown in Figure 5, is loaded from configuration files which define the coordinate locations of all the QR code post points on the map and the available paths between those locations. With the location data provided by the connection to ROS, the navigation module receives the current and previously observed locations for the robot. Using this location knowledge, and the coordinate locations of these locations, an approximate direction vector for the robot can be calculated. Next, with the current location of the robot, A* search, implemented using the `python-astar` package [36], is used to find the best path to the destination. Using the generated path, the location of the current and next locations that the robot needs to visit are used to generate a direction vector that the robot needs to follow in order to move toward the next post point on the journey. By comparing these two direction vectors, the navigation module can prepare a perception with the direction to the destination. The perception is generated and published to the `perceptions` topic, telling the robot if the destination is to the left, right, ahead, or behind. This perception is of the format `direction(DESTINATION,DIRECTION)`, where `DESTINATION` is the destination that the navigator is searching for and `DIRECTION` is either `left`, `right`, `forward`, or `behind`, or `arrived`. In the event that the destination has not been specified, both `DESTINATION` and `DIRECTION` are specified as `unknown`. This direction is finally published to `perceptions` for the agent to use in decision making.

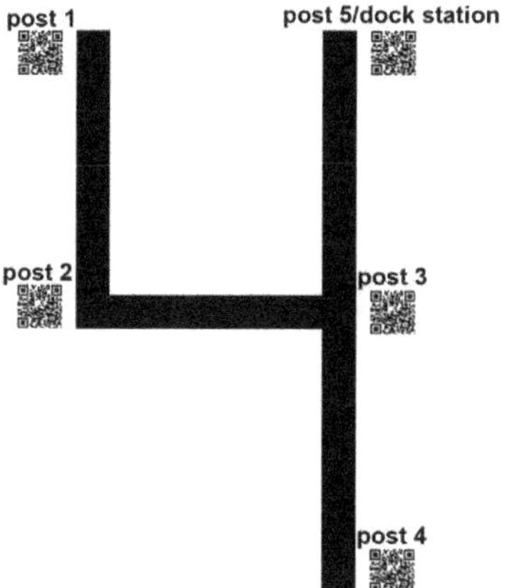

Figure 5. Map of the test environment.

5.5. User Interface

The user interface currently consists of Python script that resides on the robot, responsible for relaying messages from the user to the inbox topic and subscribing to the outbox topic. We plan to develop this into an application where a user would be able to specify specific commands for the robot using either a web browser or an Android application. On start-up, the script queries the user to specify the location of the mail sender, the mail receiver, and the docking station. The agent is informed of the docking station location using a `tell` message containing `dockStation(LOCATION)`, where `LOCATION` is the user specified location code for the docking station. The mission parameters, the sender and receiver location, are passed to the agent using an `achieve` message of the form `collectAndDeliverMail(SENDER,RECEIVER)`. This tells the agent to adopt the goal of `!collectAndDeliverMail(SENDER,RECEIVER)` with the sender location being specified by `SENDER` and the receiver location specified by `RECEIVER` [1]. Once these messages have been sent to the robot, the user interface prints messages received from the agent via the `outbox`, which provides updates of the robot's progress on its mission.

5.6. Action Translator

When the robot reasoning system requests that an action be performed by the robot, the action is published to the `actions` ROS topic. These messages are interpreted by the action translator, a Python script which subscribes to the `actions` topic and then publishes messages to the appropriate topics for the `create_autonomy` node to control the lower level hardware of the robot and to the `setDestination` topic, for setting the robot's destination in the navigation module. The action messages that are currently supported include actions for maneuvering the robot, docking and undocking the robot from a charging station, and setting the destination for the navigation module.

The maneuvering actions include `drive(DIRECTION)` and `turn(DIRECTION)`, where `DIRECTION` can be either `forward`, `left`, `right`, `stop`, or `spiral` (where the robot will drive in a widening spiral pattern). The `drive(DIRECTION)` action commands the robot to drive a short distance using the predefined motor settings for the specified direction whereas `turn(DIRECTION)` performs the drive action repeatedly until the line sensor detects that the line is centered under the line sensor. This is useful for turning at intersections, or for searching for the line if it has been lost using the spiral.

For setting the destination of the robot, the action translator sends a specified `DESTINATION` to the navigation module when `setDestination(DESTINATION)` is received. The action translator also supports actions for docking and undocking the robot from the charging station using the internal programming of the robot: `station(dock)` and `station(undock)`.

As the agent reasoning system continues the reasoning cycle once actions are sent to ROS, it is important for the action translator to ignore any conflicting actions while the action is completed. Trade-offs with respect to how actions are handled are presented in the Discussion, in Section 7.

5.7. Agent Behaviour

The reasoning system receives inputs via perceptions and a message inbox and actuates via actions and outbox messages based upon the results of its reasoning cycle. The agent behaviour is defined for the Jason BDI reasoner in AgentSpeak. The agent implementation for this project uses a hierarchy of behavioural goals, each of which have supporting plans providing the agent with a means of achieving the goals in a given context. At the top of the hierarchy are the battery charging and mail mission related goals and plans. The plans that are triggered are used to adopt sub-goals for navigating the robot to the destinations that the robot needs to visit in order to accomplish these objectives. Next in the plan hierarchy are the plans that implement the navigation behaviours. These plans are responsible for ensuring that the robotic agent travels to the required locations in the environment. Next in the hierarchy are the plans for implementing the line following behaviour, which is how the robot moves between the post points. Lastly, we have default plans for all of the goals that the agent can adopt.

We first discuss the goals associated with sending and delivering mail in Section 5.7.1. We next discuss the plans for achieving the goal of charging the battery in Section 5.7.2. Both the battery charging and mail delivery plans depend on lower level plans for navigation, discussed in Section 5.7.3. The navigation plans use the path following goals for movement between post points on the map. These goals are achieved using the path following plans discussed in Section 5.7.4. Lastly, a set of default plans are discussed in Section 5.7.5.

5.7.1. Collecting and Delivering Mail

The `!collectAndDeliverMail(SENDER,RECEIVER)` mission is the main mission of this agent. This mission is adopted by an achieve command from the user interface. For this plan, provided in Listing 2, to be applicable, the robot must not have the belief of `charging`, which would preclude the robot from being available to perform this mission. From here, the plan is very simple. First, the agent makes a mental note of the mail mission parameters, in case the intention to complete this goal, or any other goals associated with it, needs to be suspended and readopted later. Next, the agent adopts the goal of `!collectMail(SENDER)` and `!deliverMail(RECEIVER)` before finally dropping the mental note that of the mail mission parameters.

The plans for achieving the goal of `!collectMail(SENDER)` and `!deliverMail(RECEIVER)`, also provided in Listing 2, are similarly simple. First, the plan for collecting the mail for the context where the agent has not yet collected it is provided. To accomplish this, the robotic agent must first go to the sender's location and then adopt the belief that it has the mail. There is a second plan for `!collectMail(SENDER)` for the context where the mail has already been collected. This is necessary in case the agent needs to restart the mail delivery mission after being interrupted. In this case, the mail has already been collected, so no further action is necessary. The plans for `!deliverMail(RECEIVER)` are similarly implemented, instead sending the robotic agent to the receiver's location if the mail has already been collected. Otherwise, there is nothing to deliver, so no action is required.

Listing 2. Collect and deliver mail plans [33].

```
+!collectAndDeliverMail(SENDER,RECEIVER)
    :   (not charging)
    <-  +mailMission(SENDER,RECEIVER);
        !collectMail(SENDER);
        !deliverMail(RECEIVER);
        -mailMission(SENDER,RECEIVER).

+!collectMail(SENDER)
    :   not haveMail
    <-  !goTo(SENDER,1);
        +haveMail.

+!collectMail(SENDER)
    :   haveMail.

+!deliverMail(RECEIVER)
    :   haveMail
    <-  !goTo(RECEIVER,1);
        -haveMail.

+!deliverMail(RECEIVER)
    :   not haveMail.
```

5.7.2. Charging Battery

The plans which implement the battery charging behaviour are provided in Listing 3. The agent perceives the battery using three specific predicates, which are generated by the battery translator: `battery(full)`, `battery(ok)`, and `battery(low)`. Rather than having the agent adopt the goal of monitoring the battery, we instead have two plans that trigger on the addition of the predicate `battery(low)` to the agent's beliefs. For either of these plans to be applicable, the agent needs to not already be charging the battery and have knowledge of the dock station location. The first of the plans is for the context where the agent's belief base contains `mailMission(SENDER,RECEIVER)`, a predicate added to the belief base by the plans that achieve the `!collectAndDeliverMail(SENDER,RECEIVER)` goal. In this case, the agent needs to add the belief of `charging` to the belief base and drop any other intentions that the agent may have had. Next, the agent must charge the robot's battery by adopting the goal of `!chargeBattery`. Once this has been achieved, the agent can then drop the `charging` belief and adopt the goal of `!collectAndDeliverMail(SENDER,RECEIVER)` in order to finish the mail mission that was interrupted. The second plan provided for the addition of the belief `battery(low)` is applicable for the context where the agent does not have `mailMission(SENDER,RECEIVER)` in its belief base. The only difference between the body of these plans is that the agent does not need to drop intentions, nor does it need to end by adopting the goal of `!collectAndDeliverMail(SENDER,RECEIVER)`, as there was no mail mission when the battery charging plan was triggered.

There are two plans triggered by the addition of the `!chargeBattery` goal, shown in Listing 3. The first is for the context where the battery is not yet full, and the robot is not docked with the charging station. Here, we adopt the goal of `!goTo(DOCK,1)`, followed by taking the action of `station(dock)` and adopting the belief of being `docked`, to prevent this plan from executing more than once. Lastly, we readopt the goal of `!chargeBattery` for the agent to maintain the goal of `!chargeBattery` while the battery charges. The last plan in this listing is applicable for the context where the battery is fully charged, and the robot is still docked with the charging station. The plan body here is to first undock the robot and then drop the belief that the robot is docked, having successfully charged the robot's battery.

Listing 3. Battery charging plans [33].

```
+battery(low)
    :    (not charging) and
         dockStation(_) and
         mailMission(SENDER,RECEIVER)
    <-   +charging;
         .drop_all_intentions;
         !chargeBattery;
         -charging;
         !collectAndDeliverMail(SENDER,RECEIVER).

+battery(low)
    :    (not charging) and
         dockStation(_) and
         not mailMission(SENDER,RECEIVER)
    <-   +charging;
         !chargeBattery;
         -charging.

+!chargeBattery
    :    (not battery(full)) and
         dockStation(DOCK) and
         (not docked)
    <-   !goTo(DOCK,1);
         station(dock);
         +docked;
         !chargeBattery.

+!chargeBattery
    :    battery(full) and docked
    <-   station(undock);
         -docked.
```

5.7.3. Navigation Plans

The plans triggered by the addition of the goal of `!goTo(LOCATION,WATCHDOG)` are presented in Listing 4. These plans are responsible for navigating the robot to locations, called post points, on the map. These plans further adopt the goal of `!followPath` for moving between post points. This goal predicate has two parameters: the location where the robot needs to move, and a watchdog parameter. As the robot is navigating in an environment where the only means of position knowledge comes from QR codes, which are not always visible, there is a possibility that the robot may need to make a navigation decision without a visible post point code. As the navigation decisions require position knowledge, we have added a watchdog counter to help the robot assess if it is stuck in such a state. When adopting the plan to go to a new location, the watchdog parameter should be set to one.

The first plan for this goal is applicable in when the robot has not yet set a destination to navigate to. In this case, the agent needs to specify the destination for the navigation module to generate appropriate turn by turn directions. The plan body readopts this goal recursively as the robotic agent has not yet arrived at the destination. The second plan is applicable when an old destination needs to be updated to a new destination. Here, the agent has received directions from the navigation module, however the destination parameter in the associated belief is for a previously requested destination. In this case, the navigation module is updated to the new destination and the goal is readopted recursively. The third plan is for the context where the robot has arrived at the destination. Here, the robot is commanded to stop. The fourth, fifth, and sixth plans are all recursive navigation plans associated with either turning the robot around, driving forward, turning left, or turning right depending on the navigation recommendation that has been generated by the navigation module and perceived by the agent. In all of these cases, once the agent executes the necessary maneuver, the `!followPath` goal is adopted to drive the robot between post points. The last plan in this listing relates to the watchdog. If the watchdog parameter has grown past 20, meaning that the agent has tried to address this goal over 20 times, it is probable that the robot is stuck without a visible post point. In this scenario, the `!followPath` goal is adopted as well as the `!goTo()` goal, resetting the watchdog.

The effect of adopting the goal of `!followPath` is to move the robot to a location where it can see a post point. Keen readers will note that none of these plans address incrementing the watchdog. That role is completed by a default plan, discussed in Section 5.7.5.

Listing 4. Navigation plans [33].

```
+!goTo(LOCATION,_)
    :    direction(unknown,_)
    <-   setDestination(LOCATION);
         !goTo(LOCATION,1).

+!goTo(LOCATION,_)
    :    direction(OLD,_) and
         (not (OLD = LOCATION))
    <-   setDestination(LOCATION);
         !goTo(LOCATION,1).

+!goTo(LOCATION,_)
    :    direction(LOCATION,arrived)
    <-   drive(stop).

+!goTo(LOCATION,_)
    :    direction(LOCATION,behind)
    <-   turn(left);
         !followPath;
         !goTo(LOCATION,1).

+!goTo(LOCATION,_)
    :    direction(LOCATION,forward)
    <-   drive(forward);
         !followPath;
         !goTo(LOCATION,1).

+!goTo(LOCATION,_)
    :    direction(LOCATION,DIRECTION) and
         ((DIRECTION = left) | (DIRECTION = right))
    <-   turn(DIRECTION);
         !followPath;
         !goTo(LOCATION,1).

+!goTo(LOCATION,WATCHDOG)
    :    (WATCHDOG > 20)
    <-   !followPath;
         !goTo(LOCATION,1).
```

5.7.4. Path Following Plans

The plans triggered by the addition of the `!followPath` goal are responsible for the line following behaviour. The intent of this behaviour is to have the robot follow the line taped to the floor, adjusting course as needed, until a post point is visible and then stop. If the line is not visible, the agent needs to search for the line. The plans that implement this behaviour are provided in Listing 5.

First, the applicable plan for the context where there is a post point visible. In this case, there is no need to follow the path any further, so the agent stops the robot. Next, the applicable plan used for the context where the agent perceives `line(center)` and no post point is visible. Here, the robot should drive forward and readopt the `!followPath` goal. For the context where the line is lost, the agent drives in a spiral pattern in the direction that the line was last seen in an effort to search for the line, again recursively readopting the goal of `!followPath`. For the context where the line is perceived to be across, the agent uses the command to drive to the left to recenter itself over the line. Lastly, we have a plan for turning the robot to the left or to the right in order to readjust the robot over the line.

Listing 5. Path following plans [33].

```
+!followPath
    :     postPoint(A,B)
    <-    drive(stop).

+!followPath
    :     line(center) and
          (not postPoint(_,_))
    <-    drive(forward);
          !followPath.

+!followPath
    :     line(lost) and
          (not postPoint(_,_))
    <-    drive(spiral);
          !followPath.

+!followPath
    :     line(across) and (not postPoint(_,_))
    <-    drive(left);
          !followPath.

+!followPath
    :     line(DIRECTION) and
          ((DIRECTION = left) | (DIRECTION = right)) and
          (not postPoint(_,_))
    <-    drive(DIRECTION);
          !followPath.
```

5.7.5. Default Plans

It is important to have default plans for the agent if there are no other applicable plans available to achieve its goals. These may run when perceptions for sensors unrelated to the goal are received, for example when a battery perception is received on its own when the agent is working through its navigation goal. In this situation, we need to ensure that these goals are not inadvertently dropped, using recursion to readopt the goals, as necessary. The default plans used by this agent are provided in Listing 6.

The first plan ensures that the `!collectAndDeliverMail(SENDER,RECEIVER)` is not dropped in error. Next is the plan for the `!goTo(LOCATION,WATCHDOG)` goal for the context where the reasoning cycle runs on a perception other than a post point. In this scenario we readopt the goal with an increment to the watchdog `!goTo(LOCATION,WATCHDOG + 1)`. Next we have the default plans for the `!followPath` and `!chargeBattery` goals which ensure that the goal is not dropped inadvertently. Lastly, we have the default plan for `!collectMail(SENDER)` and `!deliverMail(RECEIVER)`.

Listing 6. Default plans [33].

```
+!collectAndDeliverMail(SENDER,RECEIVER)
    <-    !collectAndDeliverMail(SENDER,RECEIVER).

+!goTo(LOCATION,WATCHDOG)
    <-    !goTo(LOCATION, (WATCHDOG + 1)).

+!followPath
    <-    !followPath.

+!chargeBattery
    <-    !chargeBattery.

+!collectMail(SENDER).

+!deliverMail(RECEIVER).
```

6. Testing and Evaluation

In this section we discuss the ways that we tested at the unit level and the system level to confirm that the agent behaviour was working properly. We also discuss our performance evaluation results. The development of software for robotic systems comes with several practical challenges. Among those challenges are issues related to how to isolate and debug specific segments of the software as well as developing without necessarily having access to the actual hardware. We will discuss those issues as well as methods used to mitigate these issues that were used as part of this project. We discuss a simple AgentSpeak simulator that we developed and used in Section 6.1. Next, a custom simulated environment used for testing the higher-level behaviour of the agent is discussed in Section 6.2. System level testing of the robot in the analogue environment is presented in Section 6.3. Finally, a performance evaluation is provided in Section 6.4.

6.1. AgentSpeak Debug Tool

In writing software, it is always prudent to perform unit level tests of the various components. Behaviours programmed in AgentSpeak are no different in this regard. Unfortunately, debugging agents for robotics involves additional challenges, as it may be more difficult or impractical to isolate specific aspects of the software in the system under test. To assist in isolating specific aspects of the AgentSpeak programs, a debugging tool was developed. This tool, although very simplistic, was found to be a great asset for unit level testing and confirming that the agent behaviours were as expected under very controlled circumstances.

To accomplish this, an environment was developed for a Jason agent which reads perception inputs from a file. Each line in this file contained the perceptions meant to be sent to the agent at the beginning of each reasoning cycle. All actions that the agent takes are simply printed to the console window. When used with the Jason mind inspector debugging tool, this environment proved very useful for catching errors, especially in the plan context components as well as syntax errors. This debugging tool is available on GitHub [37].

6.2. Custom Simulator

Moving past the unit level testing discussed in the previous section, it became necessary to perform testing of the agent behaviour, as well as the other ROS nodes developed for this system, without necessarily having access to the robot and hardware. In practical work environments this can happen for several reasons. For example, the hardware and software being under development in parallel. Another possibility is that the robot is unavailable as it is in use for multiple projects, or that team members are geographically dispersed. Another reason for using simulation based testing is that testing on the robot itself may be time consuming; if every minor change in software required a time consuming experimental setup in order to test it, a developer could find themself delaying testing until there have been many new changes. The issue with such an approach is that tracing back the cause of an issue could become more difficult. Using a simulator enables the developer to test for minor changes often and find those issues. Lastly, it may also be more difficult to replicate the specific scenario needed to be tested in the real world, which may be more easily controlled in simulation.

To assist with development and testing, a mock simulation environment was developed [38]. This tool was used as a substitute environment which could be used for isolating specific aspects of the agent's behaviour. It included a grid style map of the environment with specific points on the map being designated as post points. The grid squares between those post points were the path from which the robot would be able to perceive the line. The environment also included a charging station with which the robot could dock. Based on the position and actions of the robot, the robot was able to move about this environment while perceiving sensor data from mock sensors for the battery state, line sensor and QR scanner. The battery was programmed to deplete at a configurable rate, enabling testing of the agent's battery charging behaviours.

This tool was developed as a ROS node that published to the ROS topics for the sensor data. It also monitored the `actions` topic for the agent to control it. In this setup, we were able to develop and test the perception translators, action translators, navigation module and agent behaviours without access to the robot. Despite the limitations of this test setup, including the lack of realism of the environment as well as the difficulty in assessing the performance of the hardware sensors themselves we were able to demonstrate most of the behaviours of the agent using this environment. The only behaviour that was not tested using this method was the line following behaviour as this was more easily accomplished with the robot itself. With any issues stemming from the implementation of the behaviours associated with navigation, mail delivery, and battery management resolved with the simulation, testing effort with the robot could focus on the interface with the sensors themselves and the line following behaviour. The simulation did not need to have high fidelity in order to be highly useful.

6.3. System Level Testing

A video of the robot operating in the analogue test environment is available [39]. Still images from that video are shown in Figure 6. In this case, the robot has been given the task of delivering mail collected at the top left corner of the map to a location at the bottom right corner of the map. In this example we see the robot having already collected the mail, shown in Figure 6a. Next, the robot begins to move toward the destination, detecting a post point along the path, as we see in Figure 6b. Having decided to turn, the robot continues toward the destination, shown in Figure 6c. Having received a `battery(low)` perception, the robot suspends the intention of delivering mail and instead moves toward the charging station, as is shown in Figure 6d. Figure 6e shows that the robot has docked with the charging station and is charging. Once charging is complete, the robot proceeds to the mail delivery destination, shown in Figure 6f.

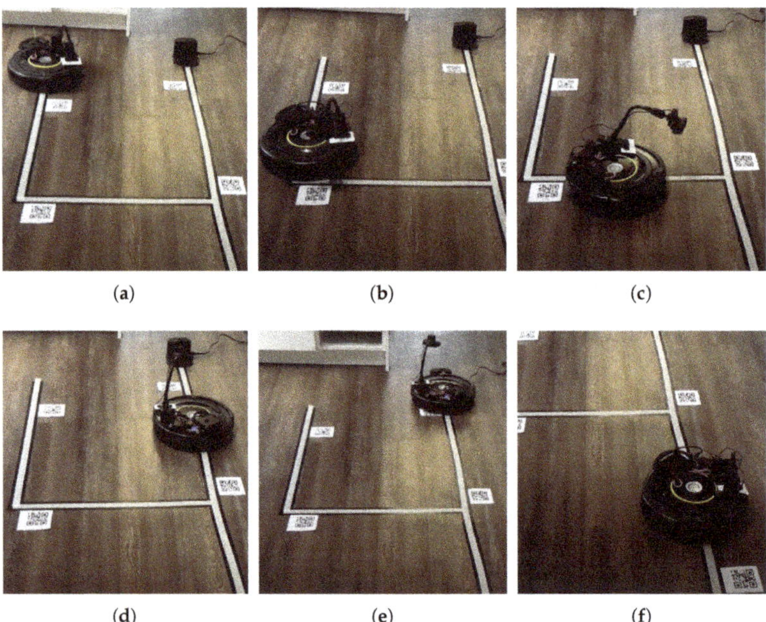

Figure 6. Demo of robot operation, video available at [39]. (**a**): mail collected; (**b**): detecting post point; (**c**): turned to continue; (**d**): interrupt delivery to charge battery; (**e**): charge battery; (**f**): proceed to destination.

6.4. Performance Evaluation

We examined the performance of the agent by logging the messages passed through the various ROS topics used by the agent. Specifically, we logged the content and time stamps of all messages passed through the `perceptions`, `actions`, `inbox`, and `outbox` topics. We also instrumented the reasoner to publish the length of the reasoning cycle to another ROS topic so that the reasoning performance could be logged. By parsing the logs of over 28 test runs we have made several observations. We will first discuss the performance of the reasoning system in Section 6.4.1 and then the plan and action frequency in Section 6.4.2.

6.4.1. Reasoning Performance

Our assessment of the performance of the reasoning system focused on whether the reasoner was able to keep up with the sensor updates. Using the logs of the perceptions, specifically their timestamps, we measured the time elapsed between publications to the `perceptions` topic resulting in a measurement of the perception period, shown in Figure 7. We also measured the time taken by the agent to perform reasoning cycles, shown in Figure 8. Outliers, which tended to be artifacts of the test start-up and shutdown process, were removed using mean absolute deviation.

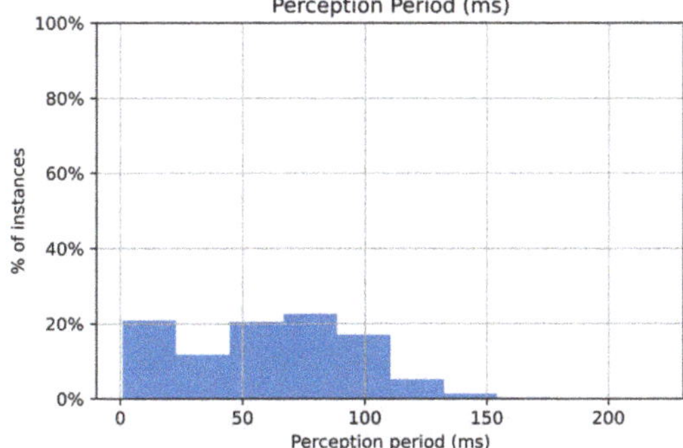

Figure 7. Perception period.

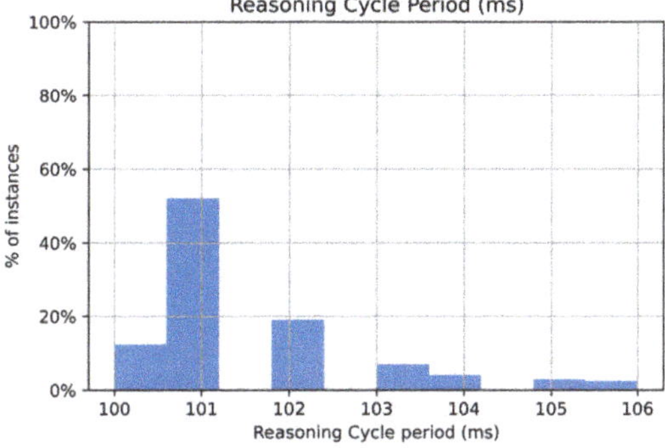

Figure 8. Processing time of the agent reasoning cycle.

We observed that the bulk of the perceptions were published more frequently than every 100 ms. By contrast we observed that the reasoning cycle generally took between 100 ms and 106 ms to complete. This means that the perceptions were usually being published at a rate that was faster than the reasoning rate meaning that the reasoning system would need to queue the perceptions as they were received. This was confirmed to be occurring by inspection of the reasoning system's perception queue. In the rare instances where the reasoning system performed faster than perceptions were received, the reasoner would wait for the next perception. What is interesting to note, however, is that the reasoning period was very reliably within a 6 ms range; the performance was consistent. Most importantly, despite this difference between the reasoning rate and the perception rate, the agent was able to properly perform its mission.

6.4.2. Plan and Action Frequency

Using logs from the `actions` and the `outbox` topic, we assessed the decision making of the agent during standard mail delivery missions. Each plan in the agent's plan base contained agent communication messages that were used for debugging purposes. These messages identified the goal that the agent was working to achieve and well as identifying information about which plan was being used. Therefore, using the `outbox` logs, we measured the proportion of time that the agent spent performing different types of plans, and for what goal it was attempting to achieve, shown in Figure 9. We also measured the proportion of the various actions that the agent performed. This is shown in Figure 10.

In examining the plan usage we see that the bulk of the time was spent achieving the `!followPath` goal, used for performing the line following task. We also see that there was usage of the default plans, specifically for the `!goTo(LOCATION,WATCHDOG)` goal as well as for `!followPath`, highlighting the importance of the default plans. Had these default plans not been provided, the agent would not have had a way of continuing the mission. We also see that, although infrequently used, the overflow plan associated with the watchdog counter for the `!goTo(LOCATION,WATCHDOG)` goal was used, validating this design choice. The `!goTo(LOCATION,WATCHDOG)` goal, and associated plans, were generally used infrequently. This was to be expected, as those plans were only to be used when navigation decisions were needed, at intersections on the map or when the robot was establishing its mission. We see that the `!collectAndDeliverMail(SENDER,RECEIVER)`, `!collectMail(SENDER)`, and `!deliverMail(RECEIVER)` goal plans were seemingly infrequently used. This was expected as the plans associated with these goals are rather short and adopt the goals of `!goTo(LOCATION,WATCHDOG)`, which in turn adopts the goal of `!followPath`. Lastly, it is important to note that, even though a plan may be infrequently used, its presence in the plan base remains essential. Without, for example, the rarely used default plans, which ensured recursion by readopting the goals, those goals would have been dropped by the reasoner and the agent would not have completed the mission.

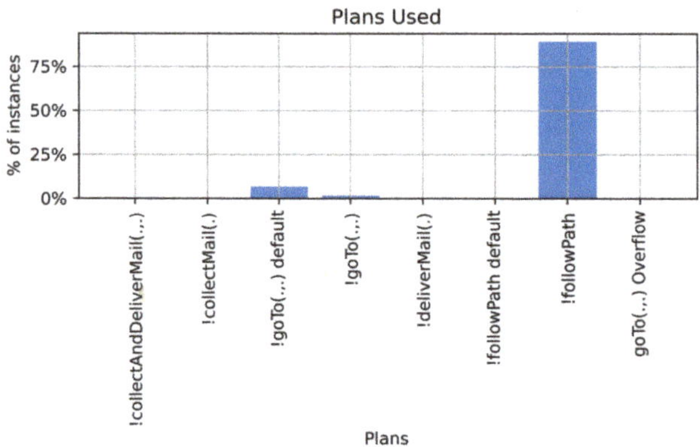

Figure 9. Plan frequency.

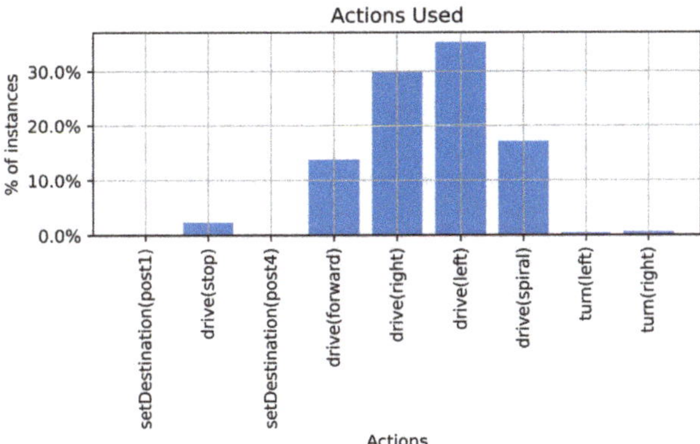

Figure 10. Action frequency.

In alignment with the plan frequencies discussed above, we see that the actions associated with the line following behaviour were the most used. We see that the plans associated with setting the destination were rare, which was expected since this occurs once per mission. We also see that the actions used for turning the robot were infrequently used as well. This was also expected as the robot only uses these actions to affect a turn at an intersection on the map. Lastly, we do see a concerning high use of the `drive(spiral)` command, which was used whenever the line sensor lost sight of the line and needed to search for the line in order to reacquire the path. This confirms our qualitative observations during testing: the line sensor was less effective than we had hoped, resulting in the robot losing the path more frequently than desired. This was especially problematic when the robot performed turns at intersections on the map.

7. Discussion

In developing agents for embedded applications, several lessons have been learned about designing such agents and the practical issues that arise in setting up such systems. Here we will discuss practical advice for developing these agents. First, we discuss plan design in Section 7.1. Next, the management of the belief base is discussed in Section 7.2. We then discuss practical issues around perceptions and actions in Sections 7.3 and 7.4, respectively.

7.1. Plan Design

In developing the behaviours for the agent discussed in this paper, various design iterations were used to find a working solution. Ultimately, we settled on a solution which uses abstraction of lower level behaviours using a hierarchy of goals and sub goals. We also defined our goals to use predicates to manage parameters. For example, for navigation we used the goal of `!goTo(LOCATION)`, where the location that the agent needs to get to is defined in the predicate. This was instead of using a belief and a generic goal, such as `destination(LOCATION)` and the goal of `!goToDestination`. Although it is possible to implement a working behaviour with both methods, the second requires that the developer manage the beliefs associated with the destination manually, whereas the first option allows the reasoning system to handle that, reducing the complexity of the plan base and ultimately reducing the likelihood of syntax errors.

Another phenomenon we encountered was that of tangled plans, where the implementation of plans for achieving a goal required intimate knowledge of the implementation of plans for other goals. We found that this was more likely to occur when revising the belief base with beliefs that were used for multiple goals. In this scenario, any updates to the plans for one goal required refactoring of the plans for the other goals as well. The goal of the developer should be that the plan implementations should be as self contained as possible, with the exception of the use of adopting goals for achieving lower level behaviour.

In planning behaviours, it is also important to consider how goals are achieved. An earlier version of the implementation of the agent used in this project used recursive plans for a goal that could not be achieved as a means of keeping the agent working on missions. Although it was possible to implement working behaviours this way, the implementation was admittedly clumsy and confusing to human readers. This can also be problematic if multiple goals that are never achieved, which are infinitely recursive, are adopted at the same time. In this scenario, the agent behaviour can become unpredictable.

The prioritization of plans in the plan base is very important. The default behaviour in Jason is that the first applicable plan in the plan base is selected as part of the intention selection function. Although this worked for our purposes, when refactoring the plan base, or simply adding and removing plans, care needs to be taken to ensure that the low priority plans are not listed too high in the implementation. For example, we opted to group all the default plans at the end of the AgentSpeak source file in order to ensure that a default plan, which was intended only to keep the agent from dropping goals prematurely, were not selected in lieu of other potentially applicable plans. Jason does allow the developer to override the event selection and intention selection functions. The authors intend to investigate these options in future work.

Consider how plans are triggered. We found that in general it made sense to make most plans trigger on the adoption of an achievement goal. We had one exception to this: the plans associated with charging the battery. These plans triggered on the perception that the battery state of charge was below a certain threshold. This was done as we needed these plans to interrupt the behaviour of the agent when potentially working toward another goal. By doing so, the agent could reason about the battery charge state when its goal was for an activity unrelated to the battery management. Had this not been implemented this way, we would have had to add context checks for the battery to plans throughout the plan base, likely tangling the plans.

Beware of death modes, situations where the agent can find itself in an unrecoverable state without any malfunctions. In the case of the mail delivery agent, a death mode existed in that the navigation plan contexts required position knowledge, which was only available when the post point QR codes were visible. If a navigation goal is adopted a time when such a perception is not available, the agent could find itself unable to execute the plan despite there being no malfunction of any components of the robot. To recover from this death mode, a watchdog was used to enable to agent to detect such modes. This timer was implemented by adding a predicate to the navigation goal which was incremented by the default plan. If the watchdog incremented past a certain threshold, the agent adopted the goal of following the path, in a hope that the robot could maneuver to a new location such that a QR code would be visible, enabling the continuation of the navigation goals.

7.2. Managing Beliefs

As mentioned earlier, it is important to manage beliefs carefully. The authors authors adopted a number of principles to facilitate this. First, if a belief is intrinsically tied to the achievement of a goal, consider refactoring the goal to use a predicate for that belief. By doing this, the developer simplifies the management of that belief in the knowledge base.

If beliefs are needed, and the option of using goal predicates is impractical, try to manage the applicable belief in the plans related only to one goal, if possible. This way, a developer does not require intimate knowledge of the implementation of plans for other goals in order to develop plans for other goals. However, if mental notes are needed for plans for multiple goals, we recommend managing these beliefs using the fewest possible number of plans in order to avoid the phenomenon of tangled plans, discussed earlier. Furthermore, be sure to remove mental notes when they are no longer needed.

7.3. Practical Management of Perceptions

Our agent implementation used ROS as a means of connecting the reasoning system to the robot hardware. A key feature of ROS is the abstraction of how nodes publish and subscribe to topics, as opposed to publishing and subscribing to other nodes directly. This means that multiple nodes can publish data relevant to perceptions at different rates: there is no guarantee that the agent will perceive data from all the sensors at every reasoning cycle. Furthermore, there is no guarantee that the sensors relevant to the goals being achieved will have been perceived at the start of every reasoning cycle. For example, the robot may be attempting to achieve the goal of `!followPath`, which primarily uses the line sensor for implementing the line following behaviour. If a reasoning cycle were to begin with the agent having only received perception data from the battery, the plan contexts associated with the line sensor would not be applicable. Therefore, it is important that a default plan be available to the agent to prevent this goal from being dropped prematurely.

Another way that this issue can manifest itself is if plan contexts use perceptions generated by different sensors. It is possible for these perceptions to be perceived in separate reasoning cycles, meaning that the desired plan context might not be applicable. If possible, the developer could avoid having plan contexts which depend on perceptions generated from multiple unrelated sensors, especially if they publish at different frequencies. Another approach could be to consider having perceptions feed into the update of the agent's beliefs about the environment. The agent's decision making could then focus on the use of these beliefs instead of the perceptions themselves. This will be explored as part of our future work.

We also found that there are scenarios where the agent works toward its goals but also receives perceptions related to the health and status of the robot. In the case of this project, the battery updates were largely irrelevant to the execution of the mail delivery mission unless the state of charge was getting too low. In this case instead of having context checks on almost all of the plans confirming that the battery state of charge was acceptable to continue, we used a high priority plan triggering on the perception of `battery(low)` which was used to adopt the goal of recharging the battery.

7.4. Practical Use of Actions

In working with our agent, we came to appreciate that there are several different types of actions. There are actions which take a short time to perform. There are actions which take a longer time to perform where the agent should wait for that action to finish. Finally, there are actions that may be more about setting a parameter that is used by some other module which performs some other required function. With this agent, all three types of actions were used.

In implementing the connection between the reasoning system and ROS, we set up the action implementation to be a publisher to the `actions` topic. Jason, however, implements actions as a function which returns a Boolean. The intention is that the action function should return `true` if the action was successful and `false` if the action was not. In our case, the function returned true if the action was successfully published to the `actions` topic. For the first type of actions, where they take a very short time to execute, for example for commanding the robot to drive a short distance, this method worked well.

For actions of the second type, which took longer to execute, for example turning at intersections, an action that took several seconds to complete, this was more problematic. With a reasoning system that executes a reasoning cycle every 100 ms or so, the agent would continue to perform reasoning before the robot had completed that action. This resulted in other actions being published, causing the robot behaviour to become erratic in these situations. A practical solution we found was to have the node which implements the actions simply ignore any conflicting actions that were published before the longer duration action was complete. Another solution would be to implement the `actions` topic with a service handler, where the service handler would return a Boolean resulting from the success or failure of the actions. This would force the agent to wait for actions to be completed prior to moving on. This would also cause the reasoning cycle period to increase, as the execution of the action would now be part of the reasoning cycle. This can also be used to ensure that action failure is more appropriately handled by the agent, which may not be able to achieve its mission if an actuator has stopped working, for example.

The third type of action was where an action sets some parameter used by another module. In the case of the mail delivery system, the action associated with setting the destination was such an action. In this case the agent needed a means of knowing that the correct destination was set, if the agent needed to change its destination. In this case, the perceptions that were generated by that node included the destination, so that the agent could confirm that the correct destination was set.

7.5. Code Readability and Troubleshooting

From a practical perspective, the authors found several useful practices which facilitated easier development and troubleshooting. Firstly, the authors found that it was prudent to avoid similarities between perceptions, knowledge, goal names, actions, etc. If these were too similar, we found that simply missing a character such as a '+' sign or an '!' would dramatically change the execution of the plans. The authors recommend adopting a naming convention when implementing their AgentSpeak programs. Secondly, it is recommended that the context guards be kept as simple as possible. Complex context guards can become difficult to read and understand and can easily become a source of error. Finally, the authors found that using agent communication could significantly help with troubleshooting and debugging. We used agent communication messages in all plans for both performance measurement and debugging purposes. These messages made it much easier to trace back what had occurred when troubleshooting.

8. Conclusions

In this paper, we presented the work to date on the development of a robotic agent for eventually performing autonomous mail delivery in a campus environment. We conclude with a description of our key accomplishments and a view toward our future work.

8.1. Key Accomplishments

We demonstrated the feasibility of using BDI in an embedded system. We accomplished this using the SAVI ROS BDI framework, linking Jason's BDI reasoning system to ROS. We also implemented our initial robot behaviours in BDI, navigating through an analogue development and testing environment using line following and QR codes while also monitoring the battery state, seeking a charging station as needed. Through performance evaluation, we observed that Jason is able to keep up with the frequency of perception updates in order to accomplish its missions, and seems therefore (at least based on this experiment) a viable language and framework for developing robot control programs that deal with multiple short-term and long-term missions.

We integrated the reasoning system onto a Raspberry Pi computer and connected it to the iRobot Create2, powering it from the robot's internal power. We integrated a line sensor and camera and developed the necessary nodes for providing their data to the BDI reasoner as perceptions via ROS. We provided a translator for the `create_autonomy` package, passing sensor data from the robot to the reasoning system, and actions back to the actuators.

8.2. Future Work

Our implementation uses line sensing and QR codes for localization in an analogue testing environment. This method was used as a first iteration for early development of our prototype system, but it has drawbacks, notably requiring the environment to have line tracks and QR codes. Additionally, the line following was more difficult that anticipated, with the robot frequently losing track of the line. One approach could be to add radio beacons to the environment and to use these as an indoor positioning system for navigation and path following. We could also add additional charging stations, which include infrared transmitters on them, for the robot to track. The robot could use these for more than just as a charging station but also as guiding beacons (make each station emit a different code to make it distinguishable by the robot), and more generally turning them into full-blown stations for mail drop-off and pick-up. These spots could be placed where wi-fi is accessible so that the robot can receive its missions and notify the recipient that the delivery is ready. We also intend to move beyond the analogue testing environment to the real world tunnel environment as our development progresses.

Another desire is to have multiple robots handling mail delivery together. The robots could work as a team, possibly handing off mail from robot to robot, and managing their battery levels. A user would not summon a specific robot to collect their mail, but would instead request the mail service, which would dispatch a robot to collect mail. From there, the robots could hand off the mail item amongst themselves while working together to deliver all mail that they have within their network. Individual robots may also carry multiple mail items. The user interface could be developed into a mobile app which can be improved to have maps of segments of the tunnel and estimates for when the mail will be delivered.

With an eye to the implementation of SAVI ROS BDI, the authors intend to investigate various design trade offs as part of their future work. This includes revisiting how actions are implemented, using ROS service handlers for implementing the actions, forcing the agent to block and wait for actions to complete, or possibly fail. The authors also intend to explore the event and intention selection functions with respect to the prioritization of plans so that the order that plans are listed in the source code is not a main factor in plan selection. Lastly, there is a need to explore knowledge and perception management for these agents. This is especially important in considering that perceptions can be generated at different rates and by different nodes, resulting in relevant perceptions not always being available at the start of applicable reasoning cycles.

Author Contributions: Conceptualization, C.O., P.G. and B.E.; methodology, P.G.; software, P.G. and C.O.; validation, P.G.; formal analysis, P.G.; investigation, P.G.; resources, C.O., P.G. and B.E.; data curation, P.G.; writing—original draft preparation, P.G., B.E. and C.O.; writing—review and editing, P.G. and B.E.; visualization, P.G.; supervision, B.E.; project administration, B.E.; funding acquisition, P.G. and B.E. All authors have read and agreed to the published version of the manuscript.

Funding: We acknowledge the support of the Natural Sciences and Engineering Research Council of Canada (NSERC), [funding reference number 518212]. Cette recherche a été financée par le Conseil de recherches en sciences naturelles et en génie du Canada (CRSNG), [numéro de référence 518212].

Acknowledgments: The team is very appreciative of the support of Simon Yacoub and Catharina Gavigan for their assistance with testing of the robot. Thank you!

Conflicts of Interest: The authors declare no conflict of interest. The funders had no role in the design of the study; in the collection, analyses, or interpretation of data; in the writing of the manuscript, or in the decision to publish the results.

Abbreviations

The following abbreviations are used in this manuscript:

AJPF	Agent Java PathFinder
AOP	Agent Oriented Programming
BDI	Belief-Desire-Intention
EBNF	Extended Backus–Naur form
GNSS	Global Navigation Satellite System
GPIO	General-Purpose Input/Output
GPS	Global Positioning System
MAS	Multi Agent System
OODA	Observe Orient Decide Act
PROFETA	Python RObotic Framework for dEsigning sTrAtegies

QR   Quick Response
ROS   Robot Operating System
SAVI   Simulated Autonomous Vehicle Infrastructure
UAV   Unmanned Aerial Vehicle

Appendix A. Additional Implementation Details

In this appendix we provide additional implementation details with respect to the robot hardware. These details are provided so that an interested reader can recreate our implementation if they so desired. First, our method of powering the computer from the robot is discussed in Appendix A.1. Appendix A.2 details the line sensor implementation.

Appendix A.1. System Power Connections

In order to power the robot's computer (a Raspberry Pi 4) without having it tethered to a socket in a wall, we utilized the iRobot Create 2's power system. This was possible as the serial connection between the computer and the robot also provides access to the robot's internal rechargeable battery. Conveniently, the serial cable used to connect the Create 2 to the Raspberry Pi exposes the robot's power bus through its RS232 pinout, as seen in Figure A1 and Table A1 [29,40].

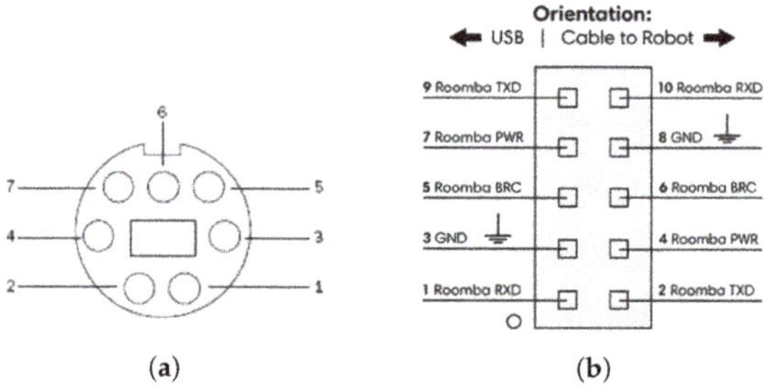

Figure A1. Serial connection pinout. (**a**): robot's RS232 pinout; (**b**): Create cable pinout.

Table A1. Create 2 external serial port RS232 connector pinout.

Pin	Name	Description
1	Vpower	Battery + (unregulated) 16 V to 20 V
2	Vpower	Battery + (unregulated) 16 V to 20 V
3	RXD	0 V to 5 V Serial input to robot
4	TXD	0 V to 5 V Serial output from robot
5	BRC	Baud rate change
6	GND	Battery ground
7	GND	Battery ground

Although this pinout provides access to the robot's power supply, it must be converted from 16 V to 20 V to the regulated 5 V required by the Raspberry Pi computer via its USB-C connector or its GPIO pin. To get a stable 5 V for our Raspberry Pi, we used a Tobsun 15 W DC to DC power converter by feeding power to its input (12 V/24 V positive and negative) terminal from pin 4 and pin 3 of the Serial to USB header described in Figure A1b respectively. We connected the exposed wires of an improvised USB type C cable to the converter and then we plugged in the cable to the Raspberry Pi; when the RS232 end of the Serial-to-USB cable is plugged into the Create2 robot, the entire system is powered successfully.

Appendix A.2. Line Sensor Implementation

As our robot operates in an indoor environment without the support of GNSS systems for navigation, a simple means of moving through the tunnels and navigation was required. As an initial implementation, a line sensor was used for the robot to follow lines on the tunnel floor. This sensor is implemented using three Photointerrupter LTH 1550-01 diodes, shown in Figure A2. Each sensor detects if the line is on the left, center or right of the robot's center. Two resistors were used per Photointerrupter, a 220 Ω and a 33 kΩ. The 220 Ω was used as a limiting resistor for the LED within the sensor and the 33 kΩ as a voltage divider to enable us to measure the voltage across the resistor when light falls on the phototransistor.

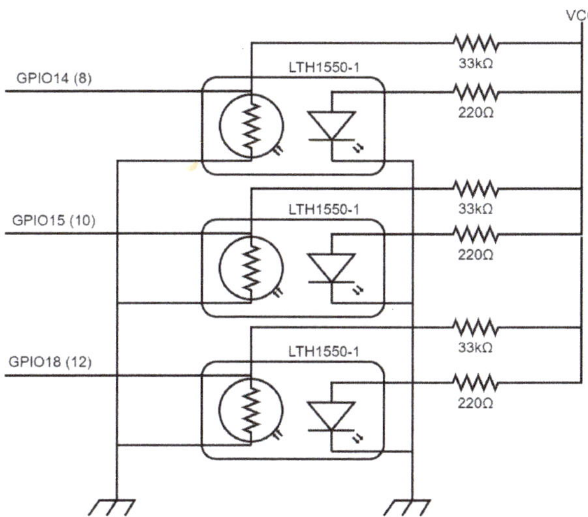

Figure A2. Line sensor circuit.

The sensors were connected to three different GPIO pins on the Raspberry Pi. The right sensor is connected to GPIO14 (pin8), the center sensor to GPIO15 (pin10) and the left sensor to GPIO18 (pin12). The sensor is powered from the Raspberry Pi; the VCC pins are connected together and then to the 5 V pin of the Raspberry Pi, while the ground (GND) pins are connected together and then to the ground (GND) pin of the Raspberry Pi. When light falls on each of these sensors, their GPIO pins are set to HIGH, and when the sensors are covered or faced with a non-reflective material or has no light falling on them, their GPIO pins are set to LOW.

The navigation track was designed using a reflective tape, so that when it is faced by any of the sensors, the respective GPIO pin is set to HIGH, and then it is known if the line is on the right, center or left depending on the pin that was set to HIGH or LOW. The sensors are mounted under the center of the Create2, in line with the right and left wheels, as seen in Figure A3a. An image of the underside of the robot itself is provided in Figure A3b. This is to ensure more navigation accuracy, because if the sensors are mounted in front, or behind the wheels, the line would be detected before or after the robot needs to make a navigation decision. For example, if the sensors are mounted in front of the wheels, and while the robot is in motion (following the line) the line changes direction; the change in direction is detected first by the sensors making the robot to turn and change its direction before it needs to, thereby making it go out of track.

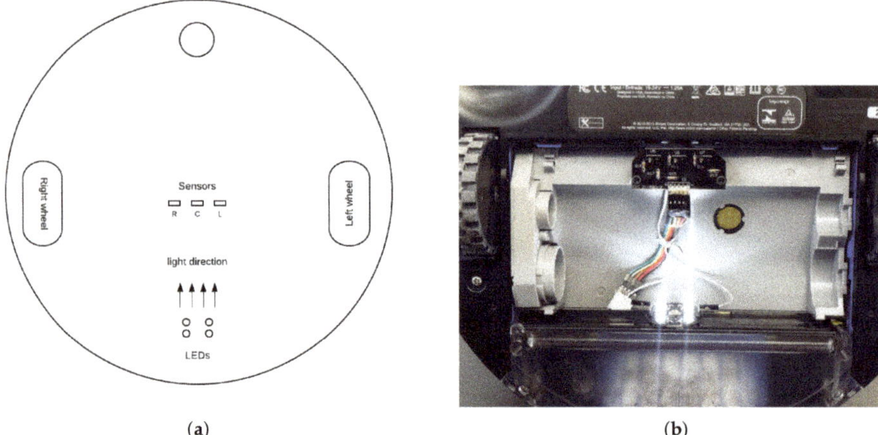

Figure A3. Layout of the underside of the robot. (**a**): robot base layout; (**b**): under the robot.

The robot has a broad surface area and when on the floor, has little or no light underneath it. Since our sensors are mounted under the robot, they cannot function effectively because they need a certain amount of light in order to detect the line. A light source under the robot using four LEDs was added. These LEDs were mounted perpendicular to the sensors with their light directed at the sensor. With this in place, when the robot is on the floor, the light bounces off any reflective object or material placed on the floor and is absorbed by non-reflective materials or objects.

The line tracks are created using tapes. To ensure enough contrast between the line to follow and the floor regardless of the environment, we had to create a track with two different types of tape, reflective and non-reflective, with the non-reflective tape in the center. This type of line track would work irrespective of location flooring. The robot is kept on the track, with the center sensor on the non-reflective tape, so when any of the sensors is faced with the non-reflective tape, we know the line track is in that direction.

References

1. Bordini, R.H.; Hübner, J.F.; Wooldridge, M. *Programming Multi-Agent Systems in AgentSpeak Using Jason*; Wiley Series in Agent Technology; John Wiley & Sons, Inc.: Hoboken, NJ, USA, 2007. [CrossRef]
2. Brooks, R. A robust layered control system for a mobile robot. *IEEE J. Robot. Autom.* **1986**, *2*, 14–23. [CrossRef]
3. Wooldridge, M. *An Introduction to MultiAgent Systems*, 2nd ed.; Wiley Publishing: Hoboken, NJ, USA, 2009.
4. Chong, H.Q.; Tan, A.H.; Ng, G.W. Integrated cognitive architectures: A survey. *Artif. Intell. Rev.* **2009**, *28*, 103–130. [CrossRef]
5. Bratman, M. *Intention, Plans, and Practical Reason*; Harvard University Press: Harvard, MA, USA, 1987; Volume 10. [CrossRef]
6. Hübner, J.F.; Bordini, R.H. Jason: A Java-Based Interpreter for an Extended Version of AgentSpeak. Available online: http://jason.sourceforge.net (accessed on 16 February 2019).
7. Open Source Robotics Foundation. ROS. Available online: https://www.ros.org/ (accessed on 27 May 2019).
8. Wallis, P.; Ronnquist, R.; Jarvis, D.; Lucas, A. The automated wingman—Using JACK intelligent agents for unmanned autonomous vehicles. In Proceedings of the IEEE Aerospace Conference, Big Sky, MT, USA, 9–16 March 2002; Volume 5, p. 5. [CrossRef]
9. Karim, S.; Heinze, C. Experiences with the Design and Implementation of an Agent-based Autonomous UAV Controller. In Proceedings of the Fourth International Joint Conference on Autonomous Agents and Multiagent Systems (AAMAS '05), Utrecht, The Netherlands, 25–29 July 2005; ACM: New York, NY, USA, 2005; pp. 19–26. [CrossRef]

10. Menegol, M.S.; Hübner, J.F.; Becker, L.B. Evaluation of Multi-agent Coordination on Embedded Systems. In *Advances in Practical Applications of Agents, Multi-Agent Systems, and Complexity: The PAAMS Collection*; Demazeau, Y., An, B., Bajo, J., Fernández-Caballero, A., Eds.; Springer International Publishing: Cham, Sweitzerland, 2018; pp. 212–223.

11. Menegol, M.S. UAVExperiments. Available online: https://github.com/msmenegol/UAVExperiments (accessed on 24 May 2019).

12. Boissier, O.; Bordini, R.H.; Hübner, J.F.; Ricci, A.; Santi, A. JaCaMo Project. Available online: http://jacamo.sourceforge.net/ (accessed on 16 May 2019).

13. Menegol, M.S. vooAgente4Wp. Available online: https://drive.google.com/file/d/0B7EcHgES6He8VEtwR 0xPZjdBbk0/view (accessed on 8 May 2019).

14. Rezende, G.; Hubner, J.F. Jason-ROS. Available online: https://github.com/jason-lang/jason-ros (accessed on 24 May 2019).

15. Rezende, G. MAS-UAV. Available online: https://github.com/Rezenders/MAS-UAV (accessed on 24 May 2019).

16. Rezende, G. jason_ros. Available online: https://github.com/jason-lang/jason_ros/ (accessed on 13 October 2019).

17. Morais, M.G. Rason. Available online: https://github.com/mgodoymorais/rason/tree/master/jason_ros (accessed on 17 July 2019).

18. Calaça, I.; Krausburg, T.; Cardoso, R.C.; Bordini, R.H. JROS. Available online: https://github.com/smart-p ucrs/JROS (accessed on 17 July 2019).

19. Fichera, L.; Messina, F.; Pappalardo, G.; Santoro, C. A Python framework for programming autonomous robots using a declarative approach. *Sci. Comput. Program.* **2017**, *139*, 36–55. [CrossRef]

20. Ledey, J.-P.; Thoreau, J. Eurobot: International Students Robotic Contest. Available online: http://www.euro bot.org/ (accessed on 15 July 2019).

21. Pantoja, C.E.; Stabile, M.F.; Lazarin, N.M.; Sichman, J.S. ARGO: An Extended Jason Architecture that Facilitates Embedded Robotic Agents Programming. In *Engineering Multi-Agent Systems*; Baldoni, M., Müller, J.P., Nunes, I., Zalila-Wenkstern, R., Eds.; Springer International Publishing: Cham, Switzerland, 2016; pp. 136–155.

22. Lazarin, N.M.; Pantoja, C.E. A robotic-agent platform for embedding software agents using raspberry pi and arduino boards. In Proceedings of the 9th Software Agents, Environments and Applications School (WESAAC), Niterói, Brazil, 1–3, June 2015.

23. Alzetta, F.; Giorgini, P. Towards a Real-Time BDI Model for ROS 2. In Proceedings of the 20th Workshop "From Objects to Agents", Parma, Italy, 26–28 June 2019; Volume 2404, pp. 1–7.

24. ElDivinCodino. ROS2BDI. Available online: https://github.com/ElDivinCodino/ROS2BDI (accessed on 29 June 2020).

25. Dennis, L.; Aitken, J.; Collenette, J.; Cucco, E.; Kamali, M.; McAree, O.; Shaukat, A.; Atkinson, K.; Gao, Y.; Veres, S.; et al. Agent-based autonomous systems and abstraction engines: Theory meets practice. In Proceedings of the 17th Annual Conference, TAROS 2016, Sheffield, UK, 26 June–1 July 2016.

26. Cardoso, R.C.; Ferrando, A.; Dennis, L.A.; Fisher, M. An Interface for Programming Verifiable Autonomous Agents in ROS. In Proceedings of the European Conference on Multi-Agent Systems (EUMAS), Thessaloniki, Greece, 14–15 September 2020.

27. Davoust, A.; Gavigan, P.; Ruiz-Martin, C.; Trabes, G.; Esfandiari, B.; Wainer, G.; James, J. An Architecture for Integrating BDI Agents with a Simulation Environment. In *Engineering Multi-Agent Systems*; Dennis, L.A., Bordini, R.H., Lespérance, Y., Eds.; Springer International Publishing: Cham, Switzerland, 2020; pp. 67–84.

28. Davoust, A.; Gavigan, P.; Ruiz-Martin, C.; Trabes, G.; Esfandiari, B.; Wainer, G.; James, J. Simulated Autonomous Vehicle Infrastructure. Available online: https://github.com/NMAI-lab/SAVI (accessed on 19 February 2019).

29. iRobot. iRobot Create2 Open Interface (OI) Specification Based on the iRobot Roomba 600. Available online: https://www.irobotweb.com/-/media/MainSite/Files/About/STEM/Create/2018-07 -19_iRobot_Roomba_600_Open_Interface_Spec.pdf (accessed on 8 March 2020).

30. Kohler, D.; Queiro, Q.; Corbellini, E. rosJava. Available online: http://wiki.ros.org/rosjava (accessed on 9 March 2020).

31.	Gavigan, P. savi_ros_bdi. Available online: https://github.com/NMAI-lab/savi_ros_bdi (accessed on 9 March 2020).

32.	Perron, J. create_autonomy. Available online: http://wiki.ros.org/create_autonomy (accessed on 8 March 2020).

33.	Gavigan, P.; Onyedinma, C. saviRoomba. Available online: https://github.com/NMAI-lab/saviRoomba (accessed on 9 May 2020).

34.	Logitech. c920-pro-hd-webcam-refresh.png. Available online: https://assets.logitech.com/assets/65478/3/c920-pro-hd-webcam-refresh.png (accessed on 9 March 2020).

35.	Onyedinma, C. SAVI_Roomba_App. Available online: https://github.com/NMAI-lab/SAVI_Roomba_App (accessed on 9 May 2020).

36.	Jrialland. Python-Astar. Available online: https://github.com/jrialland/python-astar (accessed on 24 August 2020).

37.	Gavigan, P. jasonDebuggingToolsaviRoomba. Available online: https://github.com/NMAI-lab/jasonDebuggingTool (accessed on 8 October 2020).

38.	Gavigan, P. saviRoombaDebugger. Available online: https://github.com/NMAI-lab/saviRoombaDebugger (accessed on 8 October 2020).

39.	Onyedinma, C.; Gavigan, P. Mail Delivery Robot Using BDI (Towards Campus Mail Delivery). Available online: https://youtu.be/cb63NFc24P0 (accessed on 24 August 2020).

40.	iRobo. iRobot Create Cable Pinout. Available online: https://www.irobot.com/-/media/mainsite/pdfs/about/stem/create/create2-cablepinout2018.ashx (accessed on 8 May 2019).

Publisher's Note: MDPI stays neutral with regard to jurisdictional claims in published maps and institutional affiliations.

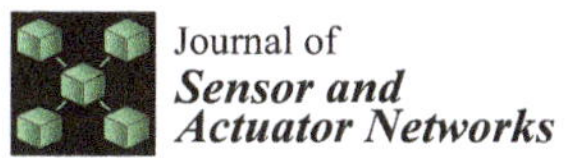

Journal of
Sensor and Actuator Networks

Article

Hybrid Verification Technique for Decision-Making of Self-Driving Vehicles

Mohammed Al-Nuaimi [1,*], Sapto Wibowo [2], Hongyang Qu [1], Jonathan Aitken [1] and Sandor Veres [1]

[1] Department of Automatic Control and Systems Engineering, The University of Sheffield, Sheffield S10 2TN, UK; h.qu@sheffield.ac.uk (H.Q.); jonathan.aitken@sheffield.ac.uk (J.A.); s.veres@sheffield.ac.uk (S.V.)

[2] Department of Electronic Engineering, State Polytechnic of Malang, Jawa Timur 65141, Indonesia; sapto.wibowo@polinema.ac.id

* Correspondence: m.al-nuaimi@sheffield.ac.uk

Abstract: The evolution of driving technology has recently progressed from active safety features and ADAS systems to fully sensor-guided autonomous driving. Bringing such a vehicle to market requires not only simulation and testing but formal verification to account for all possible traffic scenarios. A new verification approach, which combines the use of two well-known model checkers: model checker for multi-agent systems (MCMAS) and probabilistic model checker (PRISM), is presented for this purpose. The overall structure of our autonomous vehicle (AV) system consists of: (1) A perception system of sensors that feeds data into (2) a rational agent (RA) based on a belief–desire–intention (BDI) architecture, which uses a model of the environment and is connected to the RA for verification of decision-making, and (3) a feedback control systems for following a self-planned path. MCMAS is used to check the consistency and stability of the BDI agent logic during design-time. PRISM is used to provide the RA with the probability of success while it decides to take action during run-time operation. This allows the RA to select movements of the highest probability of success from several generated alternatives. This framework has been tested on a new AV software platform built using the robot operating system (ROS) and virtual reality (VR) Gazebo Simulator. It also includes a parking lot scenario to test the feasibility of this approach in a realistic environment. A practical implementation of the AV system was also carried out on the experimental testbed.

Keywords: self-driving vehicle; formal verification; model checking; rational agent; decision-making; ROS

Citation: Al-Nuaimi, M.; Wibowo, S.; Qu, H.; Aitken, J.; Veres, S. Hybrid Verification Technique for Decision-Making of Self-Driving Vehicles. *J. Sens. Actuator Netw.* **2021**, *10*, 42. https://doi.org/10.3390/jsan10030042

Academic Editors: Lei Shu, Rafael C. Cardoso, Angelo Ferrando, Daniela Briola, Claudio Menghi and Tobias Ahlbrecht

Received: 17 March 2021
Accepted: 23 June 2021
Published: 29 June 2021

Publisher's Note: MDPI stays neutral with regard to jurisdictional claims in published maps and institutional affiliations.

1. Introduction

The Defense Advanced Research Projects Agency (DARPA) sponsored competitions between 2004–2007 [1,2] presented new results on autonomous ground vehicles that showed large steps forward in the field. However, the results still primarily address non-complex driving environments [3]. AVs, which operate in complex environments, require methods that can handle unpredictable circumstances and reason in a timely manner in complex urban situations, where informed decisions require accurate perception.

The development and deployment of AVs on some of our roads are not only realistic but can also bring significant benefits. In particular, they promise to solve various problems related to: (i) the improvement of traffic congestion, (ii) the reduction of the number of accidents (iii) automate the parking operation including looking for a free parking space, and (iv) encourage shared use of AVs to reduce overall fuel consumption [4]. Studies show that more than 90% of all car accidents are caused by human errors and only 2% by vehicle failures [5].

Considerable research and development resources are spent in industry and academia on hardware and algorithms, which cover different challenges such as perception, planning, and controls. Decision-making while driving is a vital process that needs special attention.

The primary cause of human accidents comes from incorrect decisions, and there will be limited benefits in developing AVs that continue to make those incorrect decisions at a similar rate to humans. Hence, we need to make sure that any decision the vehicle is going to take has been thoroughly verified.

AVs depend on many sensors to find their way among static and dynamic obstacles; each of those sensors has strengths and weaknesses. Cameras and LiDARs are usually used together in perception systems to provide a high level of certainty. LiDAR often provides excellent odometry, localization, mapping, and range information but with a limit to object identification. Cameras provide better recognition but with limits to localization accuracy [6]. A multi-sensor system can provide reliable information for perception in joined-up software architecture for timely processing of the sensory data in the context of localization and mapping, planning, dynamic obstacle detection, and avoidance [7].

Intelligent software agents have been in development for the past two decades. Some well-known agent types are reactive, deliberative, multi-layered, and belief–desire–intention (BDI) agents [8,9]. The limited instruction set agent (LISA) [10] is a new multi-layered approach to rational agents based on the BDI agent architecture, which is particularly suitable for achieving goals by autonomous systems.

With the increasing demand for machine learning techniques and advanced planning and decision-making methods, verification and guaranteed performance of the autonomous driving process has become a challenging problem. Reconfigurable and adaptive RA-based control systems are capable of controlling a vehicle in a trajectory to avoid other vehicles and people [11]. Integration is essential to enable decision-making based on behavior rules and experience in order to make decisions with foresight and consideration to other traffic participants. RAs have demonstrated significant robustness in the implementation of various applications. However, for real-world critical applications, some safety concerns can still be raised even after extensive testing, creating the need for an appropriate verification framework. It is important to note that validation and verification usually needs to be performed together to check the system. However, this paper focuses on a new verification framework for the safety of autonomous vehicles.

The testing of systems through prototype development only answers some of the components of operational safety questions. The best that can be achieved in testing is to use a representative set of scenarios on real vehicles. Simulations can provide illustrations of the correct dynamic and social behavior of the AV. However, it is difficult to take into account rare combinations of events that may arise during the run-time of the autonomous system. It is unlikely that the designer will think of all potential scenarios to ensure complete coverage. Formal verification methods try to answer the rest of the questions by accounting for all the probabilities for a given scenario [12,13]. If accurate dynamical models are available to represent robotic skills of sensing and action, then formal verification can rely on a finite interaction model of the vehicle with a bounded model of the environment, that is based on known characteristics of traffic participants.

This paper describes a novel method for the verification of the decision-making system of an AV with a proposed architecture that lends itself to verification. We take into account the computer-based system consisting of AV design and simulation for the new verification platform. Safety and ease of implementation of the system are the two central themes in this paper, with the prime focus on the safety aspect. This paper presents a prototype system of an AV parking lot scenario with the ability to deal with the most vulnerable traffic participants: vehicles and pedestrians. In general, the level of autonomy of a vehicle can vary from fully human-operated (level 0) to a fully autonomous vehicle (level 5). Our vehicle is designed to work at level 4, where it can work autonomously in a restricted environment until it is interrupted [14].

The architecture of our proposed perception system is divided into four subsystems: LiDAR-based, vision-based, tracking-classification, and coordinate transformation. The perception system is used for localization and mapping, including calculating the relative positions of objects around the AV. The cameras are responsible for object recognition and detection of free parking spaces, with the aid of the LiDAR to provide an occupancy grid. The position of the objects is converted to the camera coordinate system, defining a region of interest (ROI) in the image space, then it obtains the depth information that belongs to that object from the LiDAR point cloud.

Most autonomous robotic agents use logic-based inference to keep themselves safe and within permitted behavior by providing the basis of reasoning for a robot's behavior [15]. Given a set of rules, it is essential that the robot can establish the consistency between its rules, its perception-based beliefs, its planned actions, and their consequences. In this paper, we are concerned with the high-level software components responsible for decisions in an AV capable of navigation, obstacle detection and avoidance, and autonomous parking. These logic-based decisions can either be implemented through a rational agent [9,10,16–19] or through fuzzy logic [20–22] depending on the level of performance guarantee required.

To achieve this, we have established the following stages. First, we have built an AV system and its environment in ROS [23] and the Gazebo Simulator [24]. Second, we investigated how a robotic agent can use model checking through the use of the MCMAS model checker [25] to examine the consistency and stability of its rules, beliefs, and actions through computational tree logic (CTL) for the RA that has been implemented within the LISA agent programming framework [10,26]. Third, we have formally specified some of the required RA properties through probabilistic timed programs (PTPs) and probabilistic computation tree logic (PCTL) formula, which are then formally verified with the PRISM Model checker [27] during run-time operation of the AV. Finally, within the proposed verification framework, which comprises both MCMAS and PRISM verification tools, we have obtained formal verification of our AV agent for some specific behaviors.

We used Gazebo Simulator in this work because of its full compatibility with ROS, and the huge support from the robotics community. PTP is a formalism for modeling systems whose behavior incorporates both probabilistic and real-time characteristics. In PTP, the location/space is discrete, while time is continuous. It is a good compromise between computational complexity and accurate mathematical modeling. Efficient verification algorithms have been developed to verify PTPs.

The development and deployment of these autonomous vehicles will rely on their situational awareness [28–32]. The vehicles will be required to co-exist alongside vehicle controlled by humans and this presents a significant problem. Whilst simulation can be used to explore edge-cases and boundaries of operation this relies on the imagination of the designer of these systems. Therefore vehicles could look to learn and adapt to situations to improve their awareness and performance. The application of this situation-based learning is out of scope for this paper but provides motivation for future work.

Contribution

This work is a continuation of our previous work [33,34] to present a new and complete verification framework for the decision-making of an AV that combines both the design-time and run-time verification. The main contribution can be summarized as follows:

1. New verification framework for decision-making of a self-driving vehicle that merges design-time verification represented by the MCMAS model checker and the run-time verification represented by the PRISM probabilistic model checker, which provide a comprehensive approach for the verification of AV's agent decisions.
2. Design, simulation, and implementation of an AV through ROS open-source physics-based system for a Tata Ace vehicle. Both the AVs in simulation and experimental implementation use the same perception, rational agent, planning, and control system software designed for a parking lot environment.

2. Related Work

Autonomous vehicles have been a major area of research interest for the research community since the DARPA Grand Challenge, which inspired the development of many AV testbeds across the industry and academia. An example is the Stanford's Junior [35], which provides a testbed with multiple sensors for planning and recognition. It is capable of dynamic object detection and tracking and also localization. Other examples are Talos from MIT [36], and Boss from CMU [37], among many others.

In this section, we discuss some recent platforms and techniques related to our work, developed for safe self-driving vehicle operation.

In ref. [38], the authors presented a testbed called cognitive and autonomous test (CAT) vehicle, which is comprised of a simulation-based self-driving vehicle, with a straight-forward transition to hardware-in-the-loop testing and execution, to support research in autonomous driving technology. The idea is to support researchers who want to demonstrate new results on self-driving vehicles but do not have an access to a physical platform to mimic the dynamics of a real vehicle in the simulation and then provide a seamless transition to the reproduction of use cases with hardware. The Gazebo Simulator utilizes ROS with a physics-based vehicle model, including simulated sensors and actuators. Gazebo comes as a default simulator with ROS and is a physics-based simulator that has also been used in our work. Gazebo is not the only option that is available to design and test self-driving vehicles. Other simulators include CARLA [39], which has been developed to support the development, training, and validation of autonomous urban driving systems. CARLA is also compatible with ROS and supports flexible specification of sensor suites and environmental conditions. Another simulator that is also useful to develop and test self-driving vehicles is LGSVL [40], which is a multi-robot AV simulator. It has been designed as an open-source simulator based on the Unity game engine to test autonomous vehicle algorithms. LGSVL also supports ROS where it helps to connect the simulator to a physical platform for tests.

Self-driving vehicles use a perception system to perceive the environment. Sensor fusion is used to bring together inputs from multiple radars, LiDARs, and cameras to form a single model or image of the environment around a vehicle. The resulting model is more accurate because it balances the strengths of the different sensors. In ref. [41], the researchers developed a perception fusion architecture based on the evidential framework to solve the detection and tracking of moving objects problem by integrating the composite representation and uncertainty management. They tested their fusion approach with a physical testbed from the interactive IP European project, which includes three main sensors: camera, LiDAR, and radar by using real data from different driving scenarios and focusing on four objects of interest: pedestrian, bike, car, and truck. The sensor fusion provides necessary information for different parts of the autonomous driving system, such as simultaneous localization and mapping (SLAM) and planning, which include both path planning and motion planning.

Other methods in the literature include the distance sensor-based parking assistance system, which recognizes an empty space using ultrasonic and LiDAR sensors as explained in refs. [42–44]. The problem with this system is that it will recognize a free space as a parking slot when the space is equal to the width of the vehicle is detected, even if the space is not a parking slot. This method is usually applied to a parking assistance system where the driver can determine a parking space. However, it is not compatible with a fully autonomous parking system, where the system judges a parking space and moves the car.

The around view monitoring (AVM) [45] can compensate for the disadvantages of distance-sensor-based detection as it can detect parking spaces based on parking slot lines instead of empty spaces. However, a false-positive (FP) can be detected from shadows and 3D objects, or the parking slot lines may be occluded by a nearby vehicle. Hence, the researchers proposed a probabilistic occupancy filter to detect parking slot lines. This filter uses a series of AVM images and onboard sensors to improve the occlusion problem and reduce the false-positive from other objects. However, this method still not very accurate and could mislead the AV in some cases.

The main topic we discuss in this paper is the verification of decision-making for our self-driving vehicle. This area of research has received more attention in the last few years as the complexity of the autonomous software has increased while the safety and feasibility of these decisions have been under investigation due to series of fatal incidents that occurred with these autonomous systems as listed in ref. [46]. In ref. [19], the authors show how formal verification can contribute to the analysis of these new self-driving vehicles. An overall representation for vehicle platooning is a multi-agent system implemented within the GWENDOLEN agent programming language in which each agent captures the "autonomous decisions" carried out by each vehicle. They used formal verification to ensure that these autonomous decision-making agents in vehicle platoons never violate any safety requirements. The authors presented a method to verify both the agent behavior using Agent Java PathFinder (AJPF) and the real-time requirement of the system using the Uppaal model checker where the system is represented as timed automata.

In ref. [47], Fernandes et al. modeled an AV with a rational agent for decision-making. To achieve this, they have established the following stages. First, the agent plans and actions have been implemented within the GWENDOLEN agent programming language. Second, they have built a simulated automotive environment in the Java language. Third, they have formally specified some of the required agent properties through LTL formulae, which are then formally verified with the AJPF verification tool. Finally, within the model checking agent programming language (MCAPL) framework they have obtained formal verification of the AV agent in terms of its specific behaviors and plans.

In ref. [48], Giaquinta et al. presented probabilistic models for autonomous agent search and retrieve missions derived from Simulink models for an unmanned aerial vehicle (UAV) and they show how probabilistic model checking using PRISM model checker has been used for optimal controller generation. They introduced a sequence of scenarios relevant to UAVs and other autonomous agents such as underwater and ground vehicles. For each scenario, they demonstrated how it can be modeled using the PRISM language, give model checking statistics and present the synthesized optimal controllers.

In our work, our system is different in the following aspects: we focused on adapting and developing different techniques and methods that contribute towards the design of a safe self-driving operation. We used ROS to design the main system functions such as the perception and control subsystems. We used a similar method presented in ref. [38] where the system built-in ROS supports hardware-in-the-loop. The difference is that our ROS system was designed to satisfy the needs for our testbed—the TATA ACE electric vehicle. Further, it provides additional functions to connect to the main decision-maker onboard and the verification system. This represents a modular overall system that supports adding more subsystems when needed.

For the perception system, we used a similar set of sensors usually used by others, where this is represented by a stereo camera, mono-cameras, and LiDAR. The combination of this set provides sufficient data to perceive the vehicle's surroundings. The perception system is presented in Section 4.

We tested our system for autonomous parking scenarios. The AV needs to look for the attached Aruco markers on each parking slot; this method is used for its simplicity, reliability, and compatibility with the fully autonomous driving mode compared with other methods mentioned in the literature. However, this method needs to be supported in the parking lot by installing Aruco markers on some or all of the parking slots to be used by AVs.

As we mentioned, this work focused on the verification of the decision-making of AVs. The novelty of our work comes from the fact that we tried to thoroughly verify the reasoner and the decisions from multiple aspects to make sure that any decision that could be made is safe to apply. We applied the verification for the reasoner offline during the design-time and online during the run-time operation.

The reasoner software has been designed by natural language programming using software called sEnglish, as explained in Section 5. This method is used for its simplicity and compatibility with the ROS system and the verification tools. However, this method comes with some limitations and it is difficult to be used with a higher-level autonomous system presented in level five autonomy.

For the design-time verification, we used the MCMAS model checker to check the consistency and stability of the logic predicates. The PRISM model checker is used to verify the decisions made by the agent during run-time operation. Details of this method are explained in Sections 6 and 7.

3. System Overview

A standard AV has a control architecture that incorporates both low-level and high-level components. The low-level components include sensors and actuators, while the high-level system components are often responsible for decision-making based on data provided by low-level components.

Our perception system provides a stream of images and 3D point cloud data obtained from sensors commonly used in AVs. It consists of eight mono-cameras (three on each side and two at the back), a stereo camera on the front, and a LiDAR on top that can be shifted left and right and tilted with a specific angle by the high-level system for better coverage. The stereo camera in front of our vehicle uses a deep-learning-based object detector that is capable of detecting different objects, including those that could exist in a parking lot environment. The perception system can also localize free parking spaces by using fiducial markers. These data are converted to high-level abstract statements that can be used by the RA onboard the vehicle. A case study of a parking lot scenario has been carried out to demonstrate the verification methods and to show the feasibility of our approach.

The AV system in Figure 1 is based on a modular design that makes practical implementations relatively simple and allows for future updates. The decision agent is central to the system design. We used the LISA agent paradigm due to its capability to execute actions based on decision-making to pursue goals while also not being too complicated to enable verification. The decision process also uses rules and abstractions from future predictions (consequences of future events) and can re-plan the path of the AV when needed.

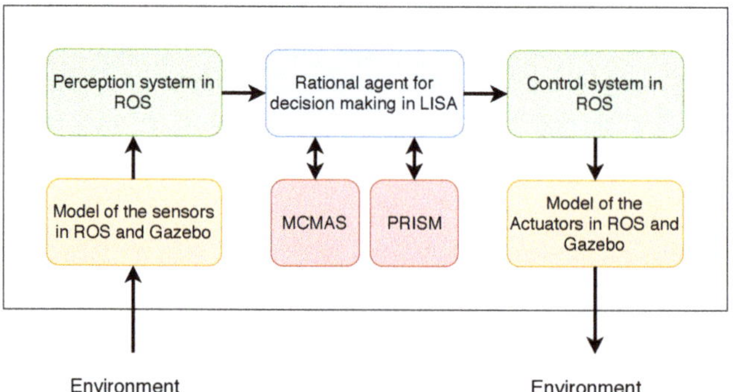

Figure 1. Block diagram showing the different components of the autonomous vehicle system in simulation. Blocks in green and yellow represent the sensing and actuating systems, the block in blue represents the rational agent that communicate with the verification system (MCMAS and PRISM).

The rational agent (RA) is capable of communicating with the perception system to sense the environment and instruct the actuators to move the vehicle in a collision-free path without the need for human support. To achieve this, the perception system builds a model of the environment, localizes objects around, and keeps updating its model after each perception cycle. The software agent has rule-based reasoning, planning capability, and some feedback control skills for steering and velocity regulation. The RA has been implemented using natural language programming (NLP) in sEnglish [17], as mentioned in Section 5.

The vehicle in simulation supports a scalable, modular design to ease the implementation of different system parts and further development. The physics-engine-based simulation shown in Figure 2 consists of a model of the Tata Ace electric vehicle shown in Figure 3, with the same specifications and parameters for the vehicle and sensors.

The AV is based on packages designed with ROS, using Python and C++. ROS provides tools and libraries for writing perception and control algorithms and other applications for AVs. With various levels of hardware and software abstractions, device drivers for a seamless interface of sensors, libraries for simulating sensors, and a visualizer for diagnostics purposes, ROS provides middleware and interoperability to simulation software, and the software installation is straightforward. Being a distributed computing environment, it implicitly handles all the communication protocols. The hardware used in this work has been selected through experimental tests, a similar set has been widely adopted by other prototypes design of AVs.

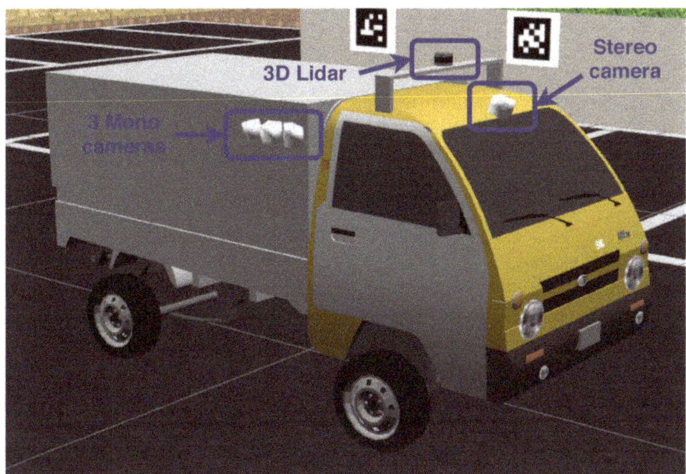

Figure 2. The test vehicle we designed in ROS and Gazebo showing sensor configuration.

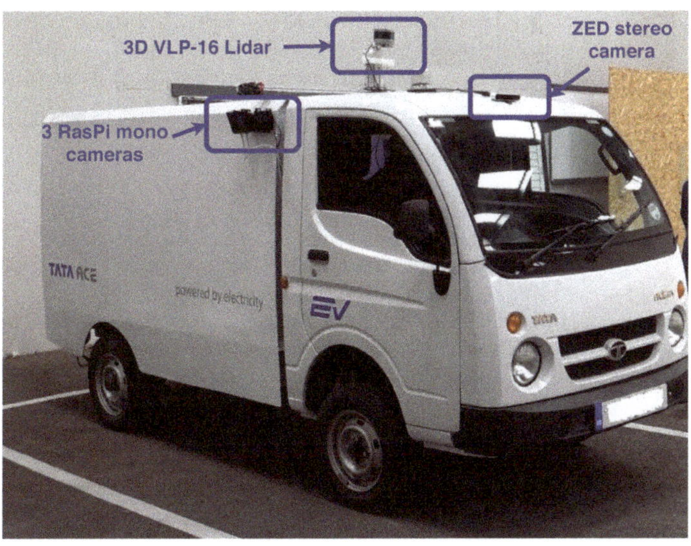

Figure 3. Our electric testbed showing the hardware and sensor configuration.

However, standard ROS packages lack domain-specific requirements for experimentation with a car-like robot. A typical setup of an AV consists of a controller and a set of sensors tested and mounted to provide sensing modality that provide a complete view of the environment around the test vehicle. In order to control motion seamlessly, we created interfaces for control and consistent consumption of sensor data. We had tested the current state of the vehicle and issued control signals well before the real platform was engaged. Once algorithms were tested in simulation, they could be implemented in the real vehicle, and the physical platform was then replaced the AV in simulation. The simulated version was used as a proving ground for the algorithms, to build confidence in their operation before transferring to the, naturally more complex, physical system.

We created models of the AV, parking lot, pedestrians, and other vehicles in the parking lot mainly using the SkechUp software [49] to create 3D models recognized by Gazebo Simulator.

In this work, we are interested in both design-time and run-time verification; this process involves the analysis of the system to detect behaviors violating the required properties. Design-time architecture verification is performed using MCMAS and probabilistic run-time verification using PRISM. We have programmed a compiler from LISA to build the models for MCMAS. The latter was then used to check the consistency and stability of beliefs, rules, and actions of the AV in its environment. When the logic predicates are inconsistent or unstable, a counterexample is generated to demonstrate the violation and help developers to correct the system [50]. PRISM is used by the RA at run-time to ask questions such as 'what is the probability of success of the current action' or 'what is the probability of achieving the current goal within a time limit' [27], the parameters used to estimate the probabilities depending on the driving scenario were, for example, the speed of the AV, speed of moving objects, and the direction of movements. For example, the agent can ask what is the probability of success if the AV were to move to a specific location within a specific period, taking into account the dynamic models (generated by the agent) of other objects moving around.

Driving in urban environments is characterized by uncertainty over the intentions and behavior of other traffic participants, which is usually considered in the behavioral layer responsible for decision-making using probabilistic planning formalisms, such as Markov decision processes (MDPs) to formulate the decision-making problem in a probabilistic framework. We used a different approach in probabilistic systems represented by probabilistic timed programs (PTPs) [51] to model the behavior of the AV and the proposed behavior of the other participants. A detailed explanation is presented in the following sections.

4. Design and Implementation of Self-Driving Vehicle

The hierarchical system is decomposed into four components, as shown in Figure 4: The perception system is used to receive information about the environment and feeds this information to the second stage. Here the agent makes decisions on the suitable progress of the car towards the destination by rules of interaction and rules of the road. The next stages are the global path planner and the local path planner, which are responsible for generating the path of the AV from the starting point to its destination based on the directions and speed profile set by the RA, then select a continuous motion plan through the environment to achieve a local navigation task. The last component is the control system that executes the motion using actuators and corrects errors in the execution in a feedback loop. In the remainder of the section, we discuss each of these components briefly.

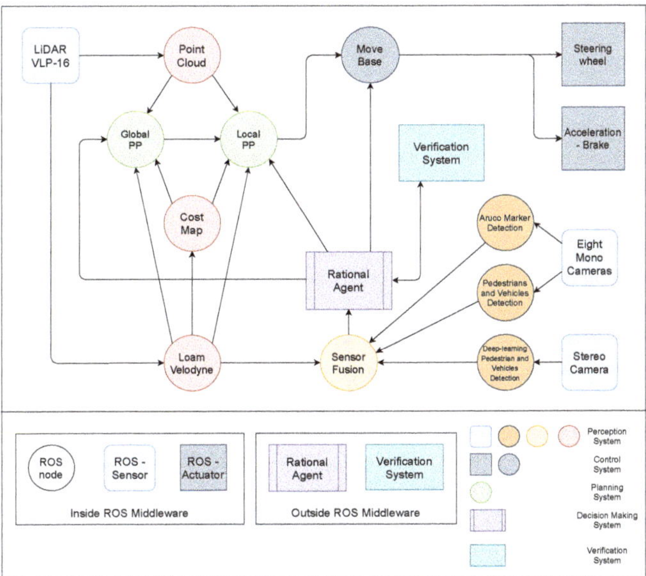

Figure 4. Our AV system showing the main nodes designed in ROS for perception, planning, and control (secondary and supporting nodes are not shown here); RA was designed in sEnglish [17,52], whereas the verification system was designed in MCMAS and PRISM verification tools.

4.1. Perception System

Modern vision-based detection techniques work by extracting image features to segment regions of interest (ROI) then detect different objects within those regions. In particular, the detection of people and vehicles has made significant progress in the autonomous and assisted driving areas [53,54]. Radar is a robust and invaluable information source for perceptual tasks; however, the spatial resolution of radar is typically poor compared to camera and LiDAR. Thus, much recent perception research is focused toward cameras and LiDAR. We have designed a parking lot scenario in ROS and Gazebo to explain the use of different sensors as shown in Figure 5.

Detection methods based on mono-cameras suffer in two ways: despite the methods proposed for moving mono-cameras, fast and accurate range measurement remains an issue, which is vital for critical object detection in autonomous driving applications. Optical sensors can suffer from a limited field of view and poor operation during low lighting conditions. On the other hand, LiDAR is usually paired with the advanced driver assistance system (ADAS) applications and has become part of the AV perception system because of the high precision range measurements and the wide field of view that it provides. The main issue for the LiDAR-based system is that the data from scans do not contain information that easily allows different objects to be distinguished between, especially in a dense environment.

Stereo cameras can provide more precise depth data and a wider angle compared with mono-cameras. However, the detection angles are smaller than LiDAR, and it is also less precise in providing depth information, especially over the long distances that are often vital for AV decision-making. The integration of cameras and LiDAR sensors can enhance fast object detection and recognition performance [41]. This type of sensor fusion system is known as the classic LiDAR-camera fusion system.

In this simulation-based system, we tried to mimic our approach for the experimental AV system, where we used a Velodyne VLP-16 LiDAR, one ZED stereo camera, and eight Raspberry Pi 3 model B mono-cameras (8 megapixels each), The properties for these sensors are mentioned in Table 1. The LiDAR is connected directly to ROS for point cloud data processing. The front-facing stereo camera is connected to the Jetson TX2 running YOLOv3 [55] deep-learning-based object detection. The mono-cameras use the processing power of their host Raspberry Pi system for aggregated channel features (ACFs) object detector of pedestrians and vehicles [56]. Those mono-cameras, along with the stereo-camera, cover a 360° FOV. The camera system has also been equipped with a method for fiducial-follow that uses Aruco markers to detect the location and orientation of free parking slots, as shown in Figure 6 (right camera one and two). Along with the occupancy grid data generated by the LiDAR, the AV is capable of detecting free parking spaces simply and efficiently. Figure 6 also shows the detection of vehicles and pedestrians, which is a vital process for normal operation of the AV.

When a known object is detected by one of the cameras, the associated LiDAR measurements are processed for the distance calculation by matching the location of the detected object with the 3D point cloud data belonging to the same object. Based on the generated depth map, the position and direction of the object are calculated from the ROI, those measurements from LiDAR are calculated according to the coordinates transformation.

We used the LiDAR odometry and mapping (LOAM) [57] ROS package for Velodyne VLP-16 3D LiDAR. This package provides a real-time method for mapping and state estimation by applying two parallel threads: The odometry thread to measure (at a higher frame rate) the motion of the LiDAR between two movements and to eliminates distortion in the point cloud. The second is a mapping thread that incrementally builds the map (at a lower frame rate) based on the undistorted point cloud, and also to compute the pose of the LiDAR on the map.

Table 1. Properties of sensors for both simulation and real testbed.

Sensor Type	No. of Sensors	Resolution	No. of Frames/Speed of Rotation
LiDAR	1	3D 16-layer (up to 50 m)-Simulation 3D 16-layer (up to 100 m)-Real testbed 360°H/30°V 1864 PPS-Simulation 300,000 PPS-Real testbed	300 RPM-Simulation 600 RPM-Real testbed
Stereo camera	1	Color 1344 × 376	10 FPS
Mono camera	8	Color 640 × 480	6 FPS

Figure 7 shows the map built for the AV current path in the parking lot shown in Figure 5. The sides of the objects that are facing the LiDAR are shown on the map with white lines. We added another layer of protection using a cost map function, which helps the AV to keep an extra safe distance from any object within a specific inflation distance where this could be set according to the environment type; it is represented on the map in Figure 7 with the blue lines surrounding the white lines. Finally, the data for the detected objects and their locations are sent to the RA for further processing.

Figure 5. Parking lot scenario developed in ROS and Gazebo Simulator to check the proposed system. The AV is looking for a free parking space in the parking lot and it is navigating among pedestrians and other vehicles depending on the data coming from the perception system and analyzed with the rational agent onboard the AV. The information obtained from the perception system is shown below in Figures 6 and 7.

Figure 6. Pedestrians and cars recognized by the AV using camera sensors. "Right camera 1" and "Right camera 2" show the Aruco marker detection attached to the parking spaces. This can be combined with the occupancy grid generated by the LiDAR in Figure 7 to detect the free parking spaces.

Figure 7. Data from the LiDAR sensor placed on the top of the AV. The odometry and mapping data are shown for the parking scenario; this is based on LOAM Velodyne ROS package shown in Figure 4. The pedestrians and the parked vehicles around have been detected and the system adds an inflation layer for extra protection from collision.

4.2. Autonomous Behavior

Recent approaches for AVs have used prediction methods in order to avoid collision by estimating the future trajectory of the surrounding traffic participants. However, real traffic scenarios include complex interactions among various road users and need to handle complex clutter and modeling interactions with other road users to ensure safety. In the DARPA urban challenge, various solutions for planning were proposed; most of those solutions were specifically tailored to the competition demands. Many approaches (e.g., refs. [36,37]) use a state machine for AV to switch between predefined behaviors. These rule-based approaches need a safety assessment in order to deal with uncertainties. AV with human-like driving behavior requires interactive and cooperative decision-making.

Other vehicle's intentions need to be modeled and integrated into a planning framework that allows for intelligent, cooperative decision-making without the need for inter-vehicle communication. While AVs need the ability to reason the intentions of other participants, those also need to infer the AV's intention reasonably. This results in inter-dependencies and interactions based on the scene and shown behavior without the need for explicit communication [58].

By simulating the proposed traffic scenario, we can search for a possible best policy measured against the AV's cost function, and then the best policy is executed from the set of available policies for the AV. Possible trajectories can also be sampled, and the reaction to the environment can be determined according to the RA model.

The AV must be able to interact with other road users in accordance with codes of conduct and road traffic rules. For a given sequence of road segments specifying the selected route, the behavior layer is responsible for selecting appropriate driving behavior based on the perceived behavior of other road users and the road conditions. For instance, when the AV searches for a vacant parking space in a parking lot, the behavioral layer instructs the vehicle to observe the behavior of other vehicles, pedestrians, and other objects during its movement and let the vehicle proceed once it is safe to go. Since the driving contexts and behaviors available in each context can be modeled as finite sets, a natural approach to automating this decision-making is to model each behavior as a state in a finite state machine with transitions controlled by the perceived driving context as the relative position to the planned route and nearby vehicles. Finite state machines, combined with different heuristics specific to the driving scenarios considered, have been adopted by most DARPA Urban Challenge teams as a behavioral control mechanism.

In the literature, similar approaches have been made by reducing decisions to a limited set of options and conducting evaluations with an individual set of policy assignments for each option. In general, the probabilistic representation of system models can be divided into four main types: discrete-time Markov chain (DTMC), continuous-time Markov chain (CTMC), Markov decision process (MDP), and probabilistic timed automata (PTA). A typical implementation of learning different driving styles in a highway simulation showed the potential of the probabilistic approach represented by MDP [59,60]. In ref. [61], the authors showed an enhanced version of the algorithm and its performance by generating human-like trajectories in parking lots, with only a few demonstrations required during learning. A partially observable MDP (POMDP), an extension of MDP, has been used in refs. [62,63] to integrate the road context and the motion intention of another vehicle in an urban road scenario.

Our RA and its physical environment have been modeled as a probabilistic timed program (PTP) model, which is an extension of PTA with the addition of discrete-valued variables that can be encoded into locations [51], while we used the probabilistic computational tree logic (PCTL) for specification logic [64]. A PTP incorporate probability, nondeterminism, dense real-time, and data. Its semantics are defined as infinite-state MDPs consists of the states of the environment and the transition between those states, which, through the conditional probabilities of the environment, correspond to triggering of predicates through the sensor system of the AV.

Pre-programmed rules are used to set the relationship between the perception predicates (beliefs) and the available actions. When this combination is verified by MCMAS during design time, then there will be no space for an unfeasible, repeated, or misleading action in the agent's actions list. The advantages of this operation are clearer when we deal with general real-life driving scenarios that could have a large number of predicates; in this case, the manual checking of those predicates would be unfeasible.

4.3. Planning System

Planning modules are concerned with vehicle motion and behavior in the perceived environment. Typically they comprise trajectory generation and reactive control for collision risk mitigation. They are organized into sub-maps to provide the flexibility of updating and to handle large environment maps. These are integrated and corrected for changes by a graph optimization approach with critical landmarks as nodes [65]. The planning system consists of two different components.

4.3.1. Path Planning

The RA is setting the waypoints to move the AV in the environment. These high-level commands are sent to the path planning ROS node to generate the route for the AV from the starting point to the desired destination. The entrance of the parking lot represents the starting point, and the destination is a free parking slot, which is an unknown place that needs to be discovered by the perception system while exploring the area. We used an ROS-based Dijkstra algorithm [66,67] for its simplicity and efficiency. This method represents the roads as a directed graph with weights represents the cost of passing a road segment. This process starts with a set of nodes (free space) that the AV can navigate and assigning a cost value to each one of them, this value is then increased with the next nodes, and the algorithm needs to find a path with minimum cost.

After finding the appropriate path, all the nodes in that path are translated into positions $P_i = (X_i, Y_i)^T$ in the reference axes. The outcome is not smooth, and some points are not compliant with the vehicle kinematics, and geometry, hence the second stage (motion planning) is necessary.

4.3.2. Motion Planning

In order to transform the global path into suitable waypoints, the timed-elastic band (TEB) motion planner creates a shorter set of waypoints $P_i = (x_i, y_i, \theta_i)^T$ within the original path planner waypoints. This takes into account, as much as possible, the vehicle constraints and the dynamic obstacles. Hence the map is reduced to the area around the AV and is continuously updating. When the path planning node determines the path of driving to be performed in the current context, then the ROS-based TEB local planner algorithm [68,69] will be used to translate this path into shorter continuous trajectories that are feasible for the control system and actuators to track and follow. This trajectory should also avoid collision with obstacles, detected by the sensors on-board, and should also be comfortable for the passengers. In case there is an object nearby, then the agent will check the possibility of collision using the PRISM model checker and then modify the trajectory when needed.

4.4. Control System

A control system is needed to execute the proposed trajectory of the AV. The move_base control node operates the AV by executing acceleration, braking, and steering messages. The control node takes the list of waypoints as input, and target velocities generated by the planning subsystem. Then sends these waypoints and velocities to an algorithm that calculates the amount of direction, acceleration, or deceleration to reach the target path.

Via a feedback controller, the appropriate actuator inputs are selected to perform the intended motion and correct the tracking of errors. These errors generated during the execution of a planned movement are due in part to the inaccuracies of the vehicle model. Therefore, the focus is placed on the robustness and stability of the controlled system. Different feedback controllers have been proposed in ref. [70] for executing the reference motions provided by the motion planning system.

Running the move_base node on the AV that is appropriately configured results in attempting to achieve a goal pose with its base to within a user-specified tolerance.

5. Rational Agent Design

5.1. Background

LISA [10] is a reactive agent that uses information from the environment in order to make a decision. These decisions are based around a set of *beliefs*, *desires*, and *intentions* that define its behavior [8,71]. Beliefs represent the knowledge derived from sensors to provide an observation of the current state of the environment. For example, if the sensors of an AV detect a person, the agent would hold the belief that a human was nearby. Desires correspond to the long-term goals of the agent. These long-term goals can correspond to states within the environment, for example, the position of an AV in a parking space, which the agent will use to attempt to establish in its behavior. Intentions, contrasting with beliefs, represent short-term goals of the agent, for example, once a person is detected, the RA will have the intention to avoid that person while they are nearby.

5.2. sEnglish

AV's decision-making programming is complicated, time-consuming, error-prone, and requires expertise in both the proposed tasks and the platform. There are many proprietary design tools in the industry [58,72] that require specialized knowledge. In order to simplify this process and to broaden the understanding of how decisions for AV actions are taken by non-experts to understand and verify the system in case of a legal need, such a method could be crucial to law enforcement agencies, insurance companies, and lawyers in the event of an accident to review the program and the reason for AV taken a specific action.

The RA is implemented using sEnglish [17,52] natural language programming. Within sEnglish, the plans operate over a description of the world, which is captured within the system and environment ontology and maintained by data from sensors in the world model. The system ontology provides a simple, translatable description between concepts that a programmer and end-user would equally understand, such as common nouns, and those that an agent can use or manipulate, such as variables or pieces of data. In sEnglish, the agent's plans are described using English sentences in a structured text, including conditioning. The meaning of sentences is explained by an sEnglish text using sentences until further decomposition of meaning reaches the signal processing level when C++ is used to define the meaning. At this C++ level, no interpretable concepts need to be defined by the ontology.

The agent takes its decisions relying on information coming from its environmental model or knowledge base, which is a database regularly updated via sensors and perception mechanisms, and potential any learned inferences. This database is organized into a high-level ontology and provides information about the system and especially the current state of the environment.

Plans are declared by the programmer. Although this makes the agent less creative at run-time, as the plan library is fixed and not dynamically generated by the agent, this has significant advantages in terms of fast execution and viable formal verification [73]. In many safety-critical systems, such as formal verification, the core agent is crucial. Hence, this kind of BDI agent combines the advantages of deliberative agents with the advantages of reliability and explainability.

5.3. Mathematical Representation of the Agent

The LISA rational agent definition of our AV will follow [9,16,17] and it is based on AgentSpeak-like BDI architectures of robotic agents.

A rational agent in LISA can be fully defined and implemented by listing the following characteristics:

- *Initial Beliefs.*
  Initially, once the agent is initialized, it will have a set of beliefs about the environment. These beliefs are referred to as $B_0 \subset \mathcal{F}$ that are a set of literals that are automatically copied into the *belief base* B_t (the set of current beliefs) on initialization.
- *Initial Actions.*
  The initial actions $A_0 \subset A$ are a set of actions that are executed when the agent is first to run. Typically these actions are general goals that activate specific initial plans set up by the programmer.
- *Logic rules.*
  A set of logic-based implication rules, $L = R^P \cup R^B$, describes *theoretical* reasoning about physics and behavior rules to enable the agent to adjust its current knowledge about the world and influence its decision on actions to be taken.
- *Executable plans.*
  A set of *executable plans* or *plan library* Π. Each plan π_j is described in the form:

$$p_j : c_j \leftarrow a_1, a_2, \ldots, a_{n_j} \tag{1}$$

where $p_j \in P_t$ is a *triggering predicate*, which allows the plan to be retrieved from the plan library whenever it comes true. Next the $p_j \in P_t$ allows the plan to be retrieved from the plan library whenever the belief base dictates that its triggering conditions are true; $c_j \in B$ is called the *context*, which allows the agent to check the state of the world, described by the current belief set B_t, before applying a particular plan; the $a_1, a_2, \ldots, a_{n_j} \in A$ then form a list of actions that the agent will execute.

The LISA rational agent defined in this paper will follow these rules and is defined:

$$\mathcal{R} = \{\mathcal{F}, B, L, \Pi, A\} \tag{2}$$

where:

- $\mathcal{F} = \{p_1, p_2, \ldots, p_{n_p}\}$ is the set of all predicates. In practice, this set can be infinitely large for general driving scenarios, however we are presenting this new approach to be tested on a specific limited driving scenario, which is driving in a parking lot. With some modifications and improvements this method could be generalized for other driving scenarios for future work.
- $B \subset \mathcal{F}$ is the atomic belief set, the set of all possible beliefs that the agent may encounter during operation. The current belief base at time t is defined as $B_t \subset B$. During operation, beliefs will always be changed. This occurs through *events* so that at a time t, beliefs may be added, deleted, or modified. These *events* are represented in the set $E_t \subset B$, which is called the *Event set*. Events may be based on *internal* or *external* actions. Internal actions are described as "mental notes". External inputs will appear through input from a sensor and are called "percepts" as they represent a measurement of the environment.
- $L = R^P \cup R^B = \{l_1, l_2, \ldots l_{n_l}\}$ is a set of implication rules. These are logic-based and represent a description of how the predicates B can be linked together and interpreted.
- $\Pi = \{\pi_1, \pi_2, \ldots, \pi_{n_\pi}\}$ is the set of executable plans or more formally *plans library*. At any time, t, there will be a collection of plans that could be activated. These are a subset of the complete plan library, $\Pi_t \subset \Pi$, which is commonly named the *Desire* set. A set $I \subset \Pi_t \subset \Pi$ of intentions is also defined. This set, l, contains plans that the agent is committed to executing. Each plan is built up as a sequence $\pi_j(\lambda_j)$ of actions where $\pi(0)$ is a triggering condition for the plan, and $\lambda_j > 0 \in A$ provides the subsequent series of actions that will be carried out.
- $A = \{a_1, a_2, \ldots, a_{n_a}\} \subset \mathcal{F} \setminus B$ is a set of all available actions. Actions may be either *internal*, when they either modify the knowledge base or generate internal events, or *external*, when they are linked to functions that operate in the environment.

This completes the definition of the AV agent used. The above list of steps are cyclically repeated to run the reasoning process of a robotic agent. Part of the agent program is shown in Figure 8 that has been used to generate PTP models for the AV and the other traffic participants based on perception predicates, the values shown are tailored to the physical characteristics of the vehicle.

In the example, the formation of plans is shown for an agent undertaking an autonomous parking maneuver. In this case, eight plans are presented that represent the agent's actions; each is represented by a triggering condition. The perception process represents sensing data that are collected on every evaluation. The '^[...]' represents the evaluation of a belief condition that can be set by an internal event. In this case, both plans start by evaluating whether a specific belief is matched. Should this belief be matched, a series of actions is then planned, again any element headed '^[...]' shows then update of a belief, elements shown within square brackets are executable sentences that contain code defined deeper within the structure which links to actuation.

Plan 1 can be read as follows: if I believe that no free parking space is detected, then I believe that I need to explore the parking lot. This is then extended by Plan 2, which can be read as if I believe I need to explore the parking lot, then a set of exploration waypoints should be generated, and these should be uploaded to activate the drive mode. Plan 3 is used to capture the condition when a parking space is detected and can be read: if I believe that I have detected a free space, then I can remove the belief that I need to explore the parking lot, and I believe I can commence parking operation.

Plan 4 contains the high-level code with trigger for planning this movement: if I believe that I can commence the parking operation, I should generate a set of waypoints for the parking and update the drive mode to reflect this. Plans 5, 6, and 7 can be read as two pairs; each deals with the detection of an object, either a person or a moving vehicle. In each case, if it is detected at a distance between 12 m and 6 m then new set of waypoints is generated to avoid the object, if the distance between 6 m and 3 m, then the drive mode is switched to a slower mode, and a new set of waypoints is generated; otherwise, the vehicle is stopped.

```
1   PERCEPTION PROCESS
2   Monitor the following Boolean:
3   Parking space located.{[],[0,5]}
4   Pedestrian detected.{[],[-2,4]}
5   Generate PTP for Pedestrian.
6   {[I am at global waypoint],[0,0]}

8   EXECUTABLE PLANS
9   //Plan 1
10  If ^ ~[Parking space located] then +^[Need to explore parking lot.].
11  //Plan 2
12  If ^[Need to explore parking lot] then [Generate exploration waypoints.
        ]
13  [Update drive mode.].
14  //Plan 3
15  If ^[Free parking lot detected] then -^[Need to explore parking lot]
16  +^[Commencing parking operation].
17  //Plan 4
18  If ^[Commencing parking operation] then [Generate parking waypoints.]
19  [Update drive mode.].
20  //Plan 5
21  If ^[Pedestrian detected] while  ^[Distance more than 3m and less that
        6m] and
22  ^[Object getting closer] then [Activate slow mode.]
23  [Generate object avoidance waypoints.]
24  +^[Object PTP generated.]
25  [Update drive mode.].
26  //Plan 6
27  If ^[pedestrian detected] while  ^[distance less than 3m] then
28  [Activate stop mode.]
29  [Update drive mode.].
30  //Plan 7
31  If ^[moving vehicle detected] while  ^[distance more than 6m and less
        that 12m] and ^[object getting closer] then
32  [Generate object avoidance waypoints.]
33  +^[object PTP generated]
34  [Update drive mode.].
35  .
36  .
37  .
```

Figure 8. Part of the agent code used to control the AV.

5.4. Connecting the RA to ROS

The sEnglish agent is natively compatible with ROS. The collection of sEnglish sentences that are set up by the programmer can comprise more complex sentences until atomic actions are then reached. These atomic actions can either be represented as sentences linked to libraries or native C++ code. The programmer can directly interface this C++ code to existing ROS libraries; therefore, the agent can be directly linked to the distributed ROS system.

A recent example of this operation is shown in handling nuclear material [18,74] for a robot arm. In this case, an sEnglish agent is developed and linked to an ROS network, in one case controlling a KUKA IIWA manipulator. In another, the agent is plugged into a different, but compatible drive for a KUKA KR180 manipulator. The only difference is the underlying drivers, providing an identical interface is provided, typically through topics and services available in ROS. The programmer can rapidly configure an sEnglish agent to operate within a distributed network for different applications.

6. Verification Methodology

In our decision framework, the agent uses model checkers MCMAS and PRISM to make appropriate and safe decisions for run-time operation. At design time, MCMAS can check if the logical reasoning system of the agent is consistent and stable [50]. The set of consistent and stable actions are fed into PRISM to find the most likely-to-succeed trajectory and action for that moment during the run-time operation.

6.1. Design-Time Verification

MCMAS is a symbolic model checker for multi-agent systems. It enables the automatic verification of specifications that use standard temporal modalities as well as the correctness, epistemic, and cooperation modalities. These additional modalities are used to capture the properties of various scenarios.

Agents can be described in MCMAS by the interpreted systems programming language (ISPL). The approach is symbolic and uses ordered binary decision diagrams (OBDDs), thereby extending standard techniques for temporal logic to other modalities distinctive of agents.

The logical reasoning system in the agent has a set of reasoning rules, which can be formulated as a Boolean evolution system (BES).

Definition 1 (Boolean evolution system). *$BES = \langle \mathcal{B}, \mathcal{R} \rangle$, where:*

- *$\mathcal{B} = \mathcal{B}^{known} \cup \mathcal{B}^{unknown}$ is a set of predicates (Boolean variables) $\mathcal{B} = \{b_1, \cdots, b_n\}$,*
- *$\mathcal{B}^{known} = \mathcal{B}^{true} \cup \mathcal{B}^{false}$ is a set of known predicates,*
- *$\mathcal{B}^{unknown}$ is a set of unknown predicates in its initial evaluation that could be determined later as $\mathcal{B}^{known}$ ($\mathcal{B}^{true} \vee \mathcal{B}^{false}$), or continue to be* Unknown,
- *$\mathcal{R}$ is a set of reasoning rules (evolution rules) of the form $X \rightarrow Y$, $\mathcal{R} = \{r_1, \cdots, r_m\}$ defined over $\mathcal{B}$.*

In the logical system, a Boolean variable in $\mathcal{B}^{known}$ usually represents a sensing event, e.g., a pedestrian comes close (e.g., within 5 m) to a vehicle. A pseudo-Boolean variable in $\mathcal{B}^{unknown}$ can express a belief, an action, or a consequence of an action, whose value is unknown at the beginning of a reasoning cycle.

When a guard g of a rule is evaluated to *true* on a valuation $\overline{B}$ of $\mathcal{B}$, we say that the rule is *enabled*. After applying all enabled evolution rules over $\overline{B}$ simultaneously, we obtain a new valuation $\overline{B}'$. If two enabled rules set a variable to different values in $\overline{B}'$, then the reasoning system is *inconsistent*. Starting from valuation $\overline{B}^0$, we can apply the evolution rules infinitely and obtain valuations $\overline{B}^1, \ldots, \overline{B}^i, \ldots$ if the reasoning system is consistent. However, the system is *unstable* if for any pair of adjacent valuations $\overline{B}^i$ and $\overline{B}^{i+1}$, we have $\overline{B}^i \neq \overline{B}^{i+1}$.

6.2. Run-Time Verification

PRISM is a probabilistic model checker [27], a verification tool for modeling and formal analysis of systems that present probabilistic behavior. PRISM has been used to analyze different kind of systems from different domains, such as planning and synthesis, communication, game theory, performance and reliability, security protocols, etc. PRISM can build and analyze several probabilistic models including Markov decision processes (MDPs) plus extensions of these models with costs and rewards.

PTPs is an extension of MDPs with real-valued clocks and state variables. For timed automata formalisms, discrete variables are typically considered to be a straightforward syntactic extension since their values can be encoded into locations.

Given a set S, $\mathcal{P}(S)$ denotes the power set of S and $\mathcal{D}(S)$ the set of discrete probability distributions over S. A PTP contains a set of state variables and a set of clock variables. The state variables model the discrete events in the environment and the clock variables model the time elapse, which is a continuous process. Let $\mathcal{X}$ be the set of *clock variables*. The set of *clock valuations* is defined as $\mathbb{R}^{\mathcal{X}}_{\geq 0} = \{t : \mathcal{X} \to \mathbb{R}_{\geq 0}\}$. Given a clock valuation t and $\delta \geq 0$, a *delayed* valuation $t + \delta$ is defined as $(t + \delta)(x) = t(x) + \delta$ for all $x \in \mathcal{X}$. Given a subset $Y \subseteq \mathcal{X}$, a new valuation $t[Y := 0]$ is defined by setting all clocks in Y to 0, i.e., $t[Y := 0](x)$ is 0 if $x \in Y$, and keeping other clocks unchanged. We used probabilistic discrete time and space in this work, hence it is necessary to use clock zones to set the time for each state and the transitions between states. A *clock zone* can be defined as a set of clock valuations that satisfy a number of clock difference constraints of the form: $\rho = \{t \in \mathbb{R}^{\mathcal{X}_0}_{\geq 0} \mid t_i - t_j \lesssim b_{ij}\}$. Let $Zones(\mathcal{X})$ be the set of all zones. Given a set $\mathcal{V}$ of state variables, let $Asrt(\mathcal{V})$, $Val(\mathcal{V})$, and $Assn(\mathcal{V})$ be a set of *assertions*, *valuations*, and *assignments* over $\mathcal{V}$, respectively.

Definition 2 (Probabilistic Timed Program (PTP) [75]**).** *A PTP is a tuple of the form:* $P = (L, l_0, \mathcal{X}, \mathcal{V}, v_i, \mathcal{I}, \mathcal{T})$ *where:*

- L *is a finite set of* locations;
- $l_0 \in L$ *is the* initial location;
- $\mathcal{V}$ *is a finite set of* state variables;
- $v_0 \in Val(\mathcal{V})$ *is the* initial valuation;
- $\mathcal{X}$ *is a finite set of* clocks;
- $\mathcal{I} : (L, \mathcal{V}) \to Zones(\mathcal{X})$ *is the* invariant condition;
- $\mathcal{T} : (L, \mathcal{V}) \to \mathcal{P}(Trans(L, \mathcal{V}, \mathcal{X}))$ *is the* probabilistic transition relation, *where:*

$$Trans(L, \mathcal{V}, \mathcal{X}) = Asrt(\mathcal{V}) \times Zones(\mathcal{X}) \times \mathcal{D}(Assn(\mathcal{V}) \times \mathcal{P}(\mathcal{X}) \times L)$$

A state of a PTP contains the valuation of L, $\mathcal{V}$, and $\mathcal{X}$, and written as (l, v, t). A new state can be reached by either an elapse of some time $\delta \in \mathbb{R}_{\geq 0}$ or a *transition* $\tau = (\mathcal{G}, \mathcal{E}, \Delta) \in \mathcal{T}(l)$ where $\mathcal{G} \in Asrt(\mathcal{V})$ is the guard, $\mathcal{E} \in Zones(\mathcal{X})$ is the enabling condition, and $\Delta = \lambda_1(f_1, r_1, l_1) + \cdots + \lambda_k(f_k, r_k, l_k))$ is a probability distribution over an update $f_j \in Assn(\mathcal{V})$, clock *resets* $r_j \subseteq \mathcal{X}$ and a *target location* $l_j \in L$.

When the agent starts a reasoning cycle, it will obtain a set of actions that can be safely applied, given the characteristics of the vehicle and measurement of the environment. This set of actions is predefined in the agent code during the design stage. If the set contains more than one action, then we use PTP to find the most suitable action for the AV to take. The most suitable action is the one that will not cause a collision, also compatible with the driving rules predefined for the agent, and will ultimately participate in reaching the destination in a shorter time and path can be considered as a safe action to apply, all of these parameters will be thought of by the agent and checked by the verification system while driving to make sure it is safe to apply. A PTP models the dynamic and uncertain physical environment containing the AV itself and other static or moving objects, such as pedestrians and other vehicles.

7. Verification of Decision-Making

This section presents an example of a parking lot scenario, where the AV is searching for a free parking space. During this process, the RA will continuously monitor the road users in its environment and decides its actions and trajectory based on the data from the perception system. The RA then checks all the probability of success of the intended actions before any execution using PRISM model checker.

MCMAS is used to verify (during design time) the beliefs, rules, actions, and their consequences that need to be considered within zone 1 and 2 of the AV, as shown in Figure 9. We used a limited set of rules and predicates for the parking lot scenario for proof of concept; real-life driving scenarios will need more rules and predicates to determine the proper behavior of the AV.

The AV needs to build a feasible trajectory and to maximize the distance from the objects around a suitable cost-map. The movements of the traffic participants are usually amenable to a probabilistic model based on the environment situation. A trajectory for a pedestrian walking in a parking lot is estimated by a prediction method [76,77], also accounting for previously collected data sets in similar scenarios, e.g., Ref. [78]. A pedestrian may keep walking at the same speed if there is a car passing nearby or could reduce the speed, stop, or change the path; the same idea can be implemented for car drivers taking into account the vehicle dynamics.

In this work, the agent generates probabilistic behavior models for the non-stationary objects based on the observed situation and from previously recorded behavior of pedestrians and drivers in real-life scenarios. The method used for trajectory prediction has been combined with prior statistics for better estimation of the object's behavior. The verification system will take into consideration probabilities for the moving objects, verifying the intended actions against them using the PRISM probabilistic model checker, to select the most likely-to-succeed action for execution. The agent keeps updating the probabilistic models of the dynamic objects and sends it to an onboard PRISM in each reasoning cycle of the agent.

This operation is repeated as long as there are no objects within zone 1 of the AV shown in Figure 9, If there is any moving object within zone 2, then the AV will halve the speed. As soon as one of the moving objects comes across zone 1, then the AV will stop based on pre-programmed rules.

7.1. Design Time Verification in MCMAS

Here we define three sets of predicates: *sensing abstractions*, *future events consequences*, and *actions*, as listed below. The operational logic of the RA is restricted to the parking lot scenario. The RA will choose its decisions based on the sensory abstractions and a set of rules, as shown in Figure 10, those rules determine the best action to be carried out by the AV based on the sensing abstractions and the possible future event consequences. MCMAS is used to compute with the resulting Boolean evolution system to verify the logical stability and consistency of those predicates.

The number of those rules could rapidly increase depending on the driving scenario and the environmental situation. While it is challenging for the designer to check that there is no conflict between them manually for this simple case study, it will be even harder when taking into account other general driving scenarios.

1. The sensory abstractions of moving objects (Zone 1/outside the trajectory of the AV) are:
   - SONO1: pedestrian detected.
   - SONO2: car detected.
   - SONO3: object detected.

2. The sensory abstractions of moving objects (Zone 1/outside the trajectory of the AV but predicted to come across) are:
   - FSNE1: pedestrian detected.
   - FSNE2: car detected.
   - FSNE3: object detected.

3. The sensory abstractions of moving objects (Zone 1/within the trajectory of the AV) are:
   - SON1: pedestrian detected.
   - SON2: car detected.
   - SON3: object detected.

4. The sensory abstractions of moving objects (Zone 2/outside the trajectory of the AV but predicted to come nearer) are:
   - FSFE1: pedestrian detected.
   - FSFE2: car detected.
   - FSFE3: object detected.

5. The sensory abstractions of moving objects (Zone 2/within the trajectory of the AV) are:
   - SOF1: pedestrian detected.
   - SOF2: car detected.
   - SOF3: object detected.
6. The sensory abstractions of moving objects moving fast (Zone 2/outside the trajectory of the AV but predicted to come nearer) are:
   - FASP1: pedestrian detected.
   - FASP2: car detected.
   - FASP3: object detected.
7. The sensory abstractions of moving objects moving fast (Zone 2/within the trajectory of the AV) are:
   - FISP1: pedestrian detected.
   - FISP2: car detected.
   - FISP3: object detected.
8. The sensory abstractions for parking the AV:
   - PSA: parking space available.
   - PSNA: parking space not available.
9. The future events (consequences) for moving objects (Zone 1) are:
   - FCN1: pedestrian detected and will be collide.
   - FCN2: car detected and will be collide.
   - FCN3: object detected and will collide.
10. The future events (consequences) for moving objects (Zone 2) are:
    - FCF1: pedestrian detected and may collide.
    - FCF2: car detected and may collide.
    - FCF3: object detected and may collide.
11. The movement actions available to AV:
    - AM1: brake to stop.
    - AM2: proceed in reduced speed (2 mph).
    - AM3: proceed in normal speed (5 mph).
12. The parking actions available to AV:
    - AA1: generate new motion plan for parking.
    - AA2: return to previous motion plan.

7.1.1. Predicates Definition

Here we define three sets of predicates: *sensing abstractions, actions,* and *future events consequences,* as listed in Section 7.1. The operational logic of the RA is restricted to the parking lot scenario. The RA will choose its decisions based on the sensory abstractions and a set of logic rules, as shown in Figure 10; those rules determine the best action to be carried out by the AV based on the sensing abstractions and the possible future event consequences. MCMAS is used to compute the resulting Boolean evolution system to verify the logical stability and consistency of those rules.

7.1.2. Worst Case Mathematical Model

Each rule can be verified by computing the minimum space-time distance of the evolution of the progress of the oncoming car/pedestrian/object (denoted by E) and that of the AV (denoted by V):

$$E : E_c + [v_e t \cos(\alpha),\ v_e t \sin(\alpha), st],\ t > t_c \tag{3}$$

$$V : V_c + [v_a t \cos(\beta),\ v_a t \sin(\beta), st],\ t > t_c \tag{4}$$

Which describes the future movements of the environmental object and the AV, respectively. t_c is the current time when sensing of E and AV decisions have been completed, and E_c and V_c are the oncoming objects and the AV position at the time of the sensor measurement are abstracted, and the decision is made by the AV what to do. We say that no collision occurs in the worst case if the geometric distance (in 3D) of these two lines is greater than 1 m for any possible heading angle α and positions E_c outside Zone 1 and Zone 2 in Figure 9. s is a time separation factor defined as $s = 1$ m/s to make the dimensions in-space time compatible and used as a scaling factor for time equivalence of space separation (the smaller s is chosen, the bigger will be the time difference requirement for two objects occurring in the same place). The validity of all rules in Figure 10 has been checked using this type of simple worst-case analysis.

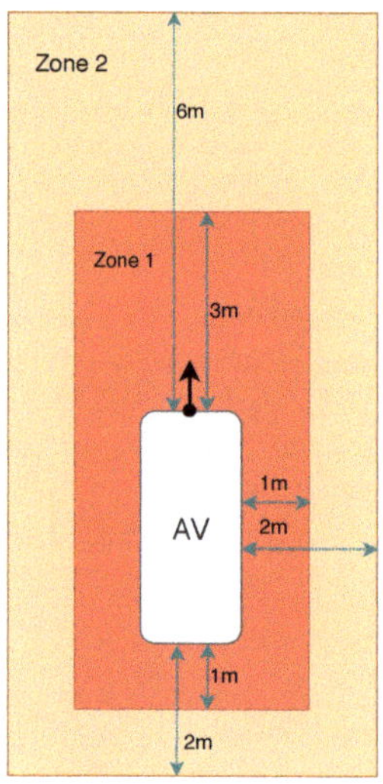

Figure 9. Schematic for the worst case analysis of the evolution rules in Figure 10. The black arrows illustrate approaching E with its sensing and decision paths extended on both sides of the 3 m and 6 m zones ahead and on the side of the AV (rectangular box).

```
56      Evolution:
57
58
59          AM1=true if FSNE1=true or FSNE2=true or FSNE3=true or SON1=true
60          or SON2=true or UON1=true or UON2=true or UON3=true;
61
62          AM1=false if FSNE1=false and FSNE2=false and FSNE3=false and
63          SON1=false and SON2=false and UON1=false and UON2=false
64          and UON3=false or AA1=true;
65
66          AM2=true if (SONO1=true or SONO2=true or SONO3=true or
67          FSFE1=true or FSFE2=true or FSFE3=true or SOF1=true or SOF2=true
68          or UOF1=true or UOF2=true or UOF3=true) and
69          (AM1=false and AA1=false and AA2=false);
70
71          AM2=false if (SONO1=false and SONO2=false and SONO3=false
72          and FSFE1=false and FSFE2=false and FSFE3=false and SOF1=false
73          and SOF2=false and UOF1=false and UOF2=false and UOF3=false) or
74          (AM1=true or AA1=true and AA2=true);
75
76      .
77      .
78      .
79      .
80      .
```

Figure 10. Sample of the agent's evolution rules in MCMAS [33].

7.1.3. Stability and Consistency Check

Computation tree logic (CTL) [79] has been used in the verification of transition systems to specify properties that a system under investigation may possess. CTL is a branching-time logic, which considers all reasonable possibilities of future behavior for our limited parking lot driving scenario. We use CTL to formulate stability and consistency checks due to the efficient implementation of CTL model checking.

CTL is given by the following grammar:

$$\varphi ::= p \mid \neg\varphi \mid \varphi \wedge \varphi \mid EX\varphi \mid EG\varphi \mid EF\varphi \mid E(\varphi\,\mathcal{U}\,\varphi) \mid$$
$$AX\varphi \mid AG\varphi \mid AF\varphi \mid A(\varphi\,\mathcal{U}\,\varphi)$$

Lemma 1. *Inconsistency in the belief base can be verified by the following CTL formula:*

$$AG(\neg(EXB_1 \wedge EX\neg B_1) \wedge \cdots \wedge \neg(EXB_n \wedge EX\neg B_n)). \tag{5}$$

Proof. If a system is inconsistent, then there must exist two successor states after a specific state such that one of them is evaluated to *true* and the other to *false*. The formula $EXB_i \wedge EX\neg B_i$ captures this case for variable b_i. The negation $\neg(\ldots)$ captures the occurrence of inconsistency through b_i. Operator AG formulates that inconsistency does not occur in any state, and we do not need to consider unknown valued variables as they cannot be assigned to unknown during evolution. $\square$

The Boolean evolution system is consistent in case the above formula evaluated to true.

Lemma 2. *The instability problem can be checked by the following CTL formula:*

$$AF((AG\ B_1 \vee AG\neg B_1 \vee AG\ K_1) \wedge \cdots \wedge$$
$$(AG\ B_{n_1} \vee AG\neg B_{n_1} \vee AG\ K_{n_1}) \wedge$$
$$(AG\ B_{n_1+1} \vee AG\neg B_{n_1+1}) \wedge \cdots \wedge$$
$$(AG\ B_n \vee AG\neg B_n)). \tag{6}$$

The Boolean evolution system is stable in case the above formula evaluated to true.

Proof. For stability, we need that every path ends with a stable state, where unknown variables will not change their values anymore. Therefore, the *unknown* variable b_i will hold one of three cases $AG\ B_i$, $AG\ \neg B_i$, or $AG\ K_i$ in the stable state. The latter means that the *known* variable cannot take value *unknown* during the evolution, and the *unknown* variables cannot take value *known*. Thus, they will not be considered in the CTL formula. AF means that eventually a stable state will be reached. $\square$

Consistent rules cannot generate contradictory conditions throughout the whole reasoning process, which means that at no time, a predicate can be assigned to *true* and *false* simultaneously. Stable rules make the reasoning process terminate in finite steps. In another word, a *stable* evaluation is reached eventually such that this stable evaluation is obtained by extending the reasoning process one step further. The detailed proofs of those lemmas are illustrated in our previous work [33,50]. Table 2 below is showing the properties of the verified system.

Table 2. Properties of the verification of the agent logic during design-time in MCMAS model checker.

Execution Time in (s)	Number of Reachable States	BDD Memory in Use	Peak Number of Nodes
0.038	1,052,670	6,641,468	16,352

7.2. Run-Time Verification in PRISM

Probabilistic decision-making and threat-assessment methods assign probabilities to different events, e.g., how likely it is to collide with another object in the next few seconds given some assumptions on uncertainties. However, when assumptions are violated (e.g., pedestrian walks/runs faster than anticipated) then sensors onboard will detect the speed of moving objects in real-time. When necessary, the vehicle will stop depending on speed measurements for the AV and other objects, and will deal with a pedestrian running towards the AV as a possible threat. The vehicle will stop if the pedestrian is within a specific distance from the driving path of the vehicle to avoid collision.

Figure 11 illustrates the proposed scenario for the AV in terms of trajectory generation based on the possible behavior of other objects around where the AV is moving forward looking for a free parking space, at the same time, a pedestrian and a vehicle (P2, V1) is moving towards the AV, another pedestrian (P1) standing in position (x = 3.5 m, y = 4 m) from the vehicle in relative coordinates. The RA will generate PTP models for the two traffic participants and also for the AV to find the best trajectory and speed under the current circumstances. The RA will keep updating the PTP models with every reasoning cycle (100 ms) and verifying those PTPs using PRISM model checker.

Because the object (P1) is not moving and it is outside (zone 2), also not in the same path of the AV, hence the agent will ignore it, and the AV will continue moving at the same speed (5 mph). However, if the pedestrian (P1) entered (zone 2), then the AV will reduce the speed according to sensory abstraction (SOF1) and action (AM2). For demonstration purposes, we discretized the trajectory by one meter apart. We also discretized the possible pedestrian's and vehicle's trajectories. While in the implementation, the RA is getting these data continuously in real-time from the perception system without the need for discretization.

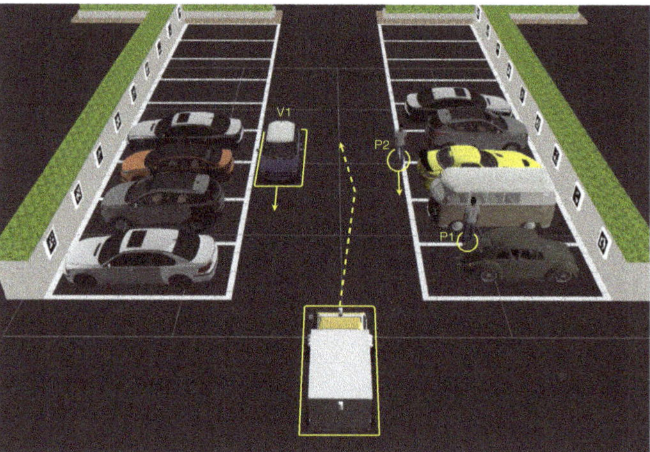

Figure 11. A driving scenario in simulation.

For the agent to build a meaningful PTP model while the AV is moving, it has been formerly equipped with a possible probabilistic behavior for both pedestrians and drivers in such an environment. Usually, when a pedestrian notices an oncoming vehicle, they may slow down with a high probability. The pedestrian may also choose to stop at some point or even change the lane to a safer one, it is also common that the pedestrian may be distracted by something, e.g., using a mobile device, and hence, does not notice the AV. If this is the case, the pedestrian continues to walk at the average speed. The last case could be included in the generated model of the traffic participants using methods explained in [80]. While there are some similar probabilities for the driver with limits to the dynamic movements of the vehicle, the driver may decide to continue driving the same speed, reduce the speed, or to stop in order to give a chance for the AV to pass easily. From the above scenario in

Figure 11, we can see that the vehicle and the pedestrian are in the same horizontal line, and this gives a small gap for the AV to pass through.

To simulate a realistic scenario, and to equip the RA with the possible behavior of pedestrians and drivers, we recorded some data and objects behavior manually for parking lots, and we used JAAD dataset [81] for pedestrians and drivers reactions to vehicles around them in different scenarios. A next step implementation would be to equip the agent with probabilistic behavior prediction method, e.g., Refs. [82,83] instead of the limited approach adopted in this work.

Generating PRISM Models from the Agent Code

We designed a translator that works as an Eclipse plugin (part of the sEnglish system environment) to translate the agent reasoning code to PTP models that can be verified by PRISM; it is a direct text processing algorithm in C++ that can run in a few milliseconds (hence its time is neglected). The agent will also translate the properties of the models in PCTL and the query of questions the agent needs to ask. As soon as the equivalent PTP models verified, then the agent will know about different properties expressed in PCTL. A Boolean variable for each belief is defined, and transition probabilities are taken from the probability distributions defined in the sEnglish code.

7.3. Verification Example of a Parking Scenario

As mentioned before, all the possible states of the system can be explored during formal verification, including some extreme cases that may be difficult to discover during testing. A general parking scenario will be presented here to illustrate the use of the RA predicates (sensory abstractions, beliefs, actions, and future event consequences) designed for this case study: we have defined two regions around the AV for safety purposes depending on the direction of movement, as shown in Figure 9, assuming the AV is moving forward, as soon as the AV detect an object within (6 m) in front or (2 m) any other side represented by (zone 2), the AV will slow down from average speed of (5 mph) to (2 mph), as soon as this object become within (3 m) from the front of the AV or (1 m) from any other side represented by (zone 1), the AV will stop. Here it is essential to mention that the experimental AV has been equipped with a means of communication with other pedestrians and drivers using audio to prevent a deadlock state when the AV stop and wait for others to move and vice versa, the AV will play a voice to say to others that "you are free to move and the AV will wait for you". We have defined further details for the AV to deal with the traffic participants around by calculating the speed of those objects using the LiDAR sensor. Assuming there is an object moving fast towards the AV, as soon as this object enter (zone 2) the AV will stop instead of slowing down, this will give more time for the other object (running pedestrian or fast-moving car) to reduce their speed, change direction, or to stop, and this will reduce or eliminate any possible collision.

A simple proposed example of how the agent chooses its actions is as follows: based on the scenario in Figure 11, a car (V1) is coming in the opposite direction to the AV from a distance of (8 m) and the driver starts to slow down when they notice another vehicle coming, the AV is moving at its average speed and building its trajectory based on the map and the moving objects around. As soon as the other vehicle (V1) enters (zone 2), the sensing event (SOF2) from Section 7.1 will be activated, and this will activate the future event (FCF2) then this will trigger action (AM2), which leads to slow down. In the meanwhile, the walking pedestrian (P2) enters (zone 2), sensing event (FSFE1) is triggered and this may lead to collision according to a future event (FCF1), the AV here will not take any further action because it is already working in reduced speed. However, as soon as the car or the pedestrian or both enter (zone 1) (SON2 or FCNE1 or both), future events (FCN1 or FCN2, or both), this will trigger the action (AM1) to execute stop action. All the stationary parked cars in the parking lot will not be considered as a threat because they are not in the proposed path of the AV, and they are not moving.

The regulation for the speed of vehicles in a parking lot is limited to (10 mph), based on this and for safety reasons and prototype development we set the speed of our AV to be (5 mph) in case of no moving objects within (zone 1 and 2). As mentioned, both the RA and the planning system will send control commands to the move base system to set the movements of the AV. However, actions such as (AA1 and AA2) have a pre-programmed sequence for performing a parking maneuver, as shown in the video link we referred to in the abstract. In case there are two or more rules in conflict with each other, MCMAS will present this case by a counterexample showing how the inconsistency is reached. Further, it cannot be the case that two different actions are activated at the same time.

For the run-time verification, the initial PTP model generated by the RA for the AV's trajectory is shown in Figure 12. We use a relative coordinate system considering that the LiDAR position on the top of the AV is the center of the coordinates at any time, knowing that the RA is taking the dimensions of the AV into calculations while processing. In this example we will refer to the coordinates of the participants according to a fixed moment at a particular time interval (x_1, y_1) to represent the coordinates of the AV, (x_2, y_2) for object (P2), the (x_3, y_3) for object (V1). The (C) letter in the PTP models represents the clock, which will be counting and resetting with every transition. The complexity of solving PTPs with two or more clocks is EXPTIME-complete. Our previous work [64] shows experiments on several complex models and properties and the results are promising.

Figure 13 shows the PTP model for the pedestrian's possible behavior. For this example, we assume that the average speed of the pedestrian is near the speed of the AV inside the parking lot. The pedestrian may prefer to stop after noticing the AV with the probability of (0.1) or to stop later when the distance became critical. We assume that the pedestrian will keep walking in the same lane with a probability of (0.6), they could also decide to change the lane and walk behind the moving car (V1) for more safety with a probability of (0.3). In both cases, the pedestrian may prefer to walk at the same speed or to reduce it with some probabilities, as shown in Figure 13.

Figure 14 shows the possible PTP model for (V1). We assumed that the driver might notice the AV and decide to stop with probability (0.1). With a probability of (0.6), the driver may decide to slow down, or may prefer to continue the same speed with a probability of (0.3). The RA will then modify the AV's PTP model according to the newly generated behavior model of the other objects around.

Note that the parameters used to generate the PTP models, such as the speed and probability, may not reflect the exact behavior of the AV, P2, or V1. The RA is building those PTPs based on the location, speed, and direction of the moving objects. In general, this framework will help to predict a possible behavior for the different objects around, then to verify the current trajectory/action for the AV against the possible trajectory/action of the nearby objects, and this will help in reducing the possibility of collision. More accurate PTP models could be generated after collecting more behaving data through real tests.

To avoid any possible collision, we require that the pedestrian and/or the vehicle is at least (1 m) away from the AV. This can be represented by the following expression:

$$\phi \equiv (x_1 - x_2)^2 + (y_1 - y_2)^2 >= 1. \tag{7}$$

$$\phi \equiv (x_1 - x_3)^2 + (y_1 - y_3)^2 >= 1. \tag{8}$$

where (x_2, y_2), represent the coordinate of the pedestrian, (x_3, y_3) is the coordinate for the car. As PRISM cannot deal with real numbers, we multiply the distance by 2 (we partition the distance by 0.5 m. Therefore, the location would have values such 0.5 and 1.5 m, by multiplying it by 2, we obtain an integer). We compute the maximum probability of the violation of Equations (7) and (8), by the following PCTL property:

$$P_{max=?}[F \neg\phi]. \tag{9}$$

Due to the discretization of the trajectory, the negation of Equation (7) is translated into the following expression:

$$(((x_2 > x_1 \land x_2 - x_1 \leq 1) \lor (x_1 > x_2 \land x_1 - x_2 \leq 1)) \land$$
$$((y_2 > y_1 \land y_2 - y_1 \leq 1) \lor (y_1 > y_2 \land y_1 - y_2 \leq 1)))$$

While the negation of Equation (8) is translated into:

$$(((x_3 > x_1 \land x_3 - x_1 \leq 1) \lor (x_1 > x_3 \land x_1 - x_3 \leq 1)) \land$$
$$((y_3 > y1 \land y_3 - y_1 \leq 1) \lor (y_1 > y_3 \land y_1 - y_3 \leq 1)))$$

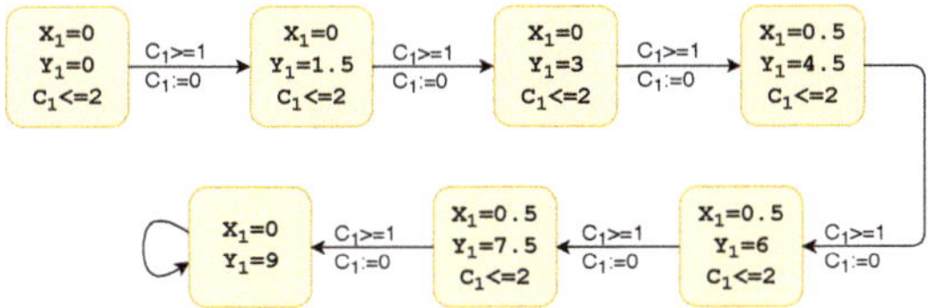

Figure 12. Initial PTP model for the AV's behavior.

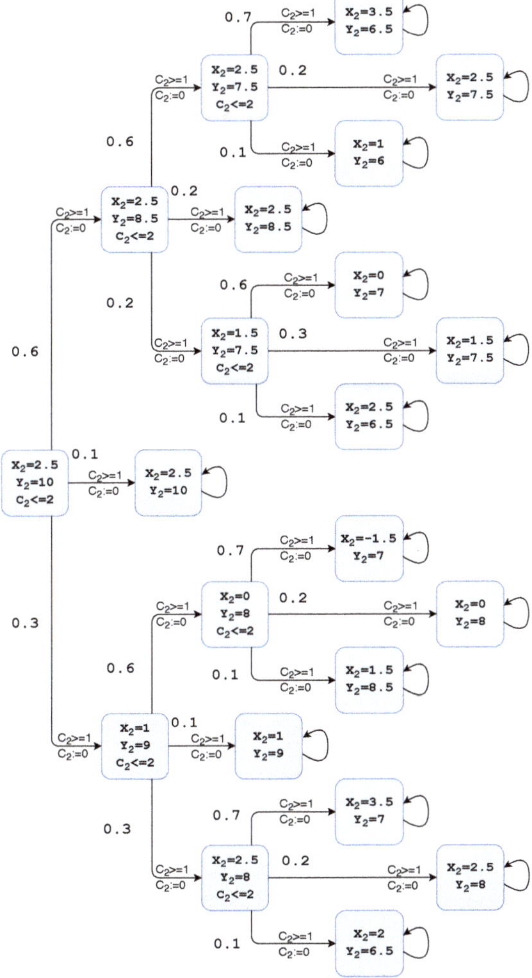

Figure 13. PTP model for the pedestrian's behavior.

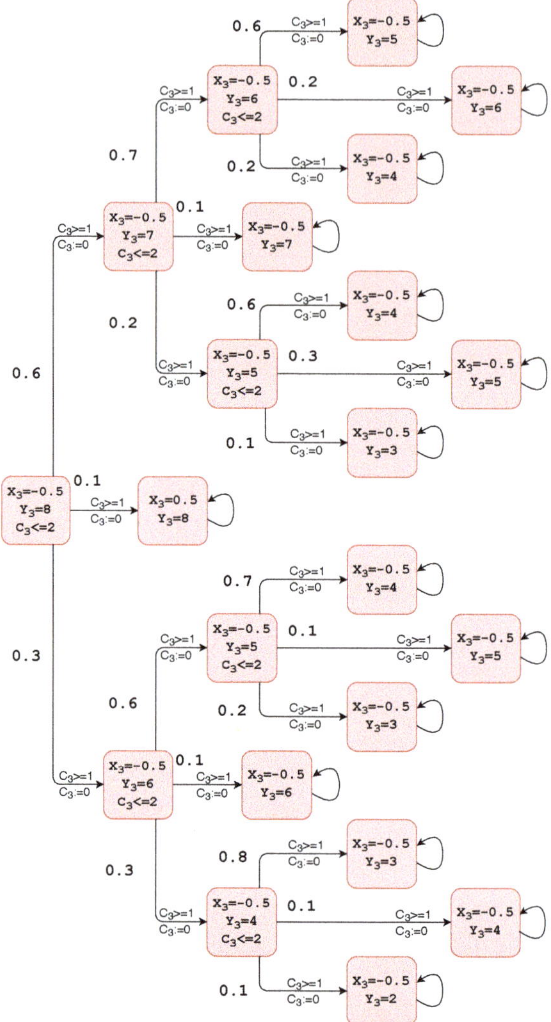

Figure 14. PTP model for the vehicle's behavior.

The verification results for the proposed scenario are shown in Table 3 returned from PRISM for Formula (9), which indicates information about the model generated for both the pedestrian and the car and the chance of collision with every one of them under the current motion plan.

Table 3. Verification results for the proposed scenario using PRISM model checker.

PTP Model	States	Transitions	Choices	Ver. Time	Maximum Collision Probability
Pedestrian	1238	3884	3624	0.036 s	0.252
Car	659	2230	1960	0.019 s	0.003

All the computations in this work were carried out using two computers running on Ubuntu OS version 16.04, first is equipped with (Intel core i7 CPU, 16 GB of RAM, and GTX 1070 GPU) for simulation, perception, planning, and control systems and the second with (Intel core i7, 16 GB of RAM, and GTX 860 GPU) to run the agent code and the verification platform.

8. Conclusions and Future Work

A new approach is presented for the verification of an agent-based decision-making system for a self-driving vehicle. The approach considers both the design-time and run-time verification. To contribute towards the open-source development of the self-driving vehicles, a self-driving vehicle is presented in the simulation that is available in ROS and the Gazebo Simulator.

A rational agent in a real traffic scenario usually faces a vast amount of situations with related behavior rules. Many of these can be identified during the design stage. Remaining scenarios, with possible probabilistic events in the environment, can then be handled by run-time evaluations. Our approach is presented through a case study. The power of the combination of the two verification tools can help the designer to eliminate any conflict and redundancy in the agent predicates. Further, the verification tools can help to check the agent rules for any possible instability or inconsistency with the benefit of obtaining a counterexample when a faulty state has been reached.

The second stage of verification deals with the possible behavior of the traffic participants to determine the probability of success for the best AV action. A limited set of beliefs, rules, and actions are presented to provide a proof of concept and to illustrate the proposed platform. For higher levels of rationality, the agent could yet be equipped, during design time, with a methodology for rules and predicates generation. Such a system would be able to learn new driving scenarios for run-time verification by implementing a machine learning approach.

Author Contributions: Conceptualization, M.A.-N. and S.V.; methodology, M.A.-N.; software, M.A.-N. and S.W.; validation, M.A.-N.; formal analysis, M.A.-N. and H.Q.; investigation, M.A.-N. and S.W.; resources, M.A.-N. and J.A.; data curation, M.A.-N.; writing—original draft preparation, M.A.-N., H.Q. and J.A.; writing—review and editing, M.A.-N., H.Q. and J.A.; visualization, M.A.-N., H.Q. and J.A.; supervision, S.V.; project administration, M.A.-N. and S.V. All authors have read and agreed to the published version of the manuscript.

Funding: This research received no external funding.

Informed Consent Statement: Not applicable.

Data Availability Statement: Data available on request due to restrictions. The data presented in this study are available on request from the corresponding author. The data are not publicly available due to privacy.

Acknowledgments: A special thanks to: (1) TATA Motors European technical center for providing the electric vehicle and some sensors used in this work. (2) AMRC research center for hosting the researcher and the vehicle and for providing a space for tests. (3) Mark Tucker and Maradona Rodrigues, TATA Motors, for their valuable comments and suggestions on this project.

Conflicts of Interest: The authors declare no conflict of interest.

References

1. Buehler, M.; Iagnemma, K.; Singh, S. *The 2005 DARPA Grand Challenge: The Great Robot Race*; Springer: Berlin/Heidelberg, Germany, 2007; Volume 36.
2. Buehler, M.; Iagnemma, K.; Singh, S. *The DARPA Urban Challenge: Autonomous Vehicles in City Traffic*; Springer: Berlin/Heidelberg, Germany, 2009; Volume 56.
3. Furgale, P.; Schwesinger, U.; Rufli, M.; Derendarz, W.; Grimmett, H.; Mühlfellner, P.; Wonneberger, S.; Timpner, J.; Rottmann, S.; Li, B.; et al. Toward automated driving in cities using close-to-market sensors: An overview of the v-charge project. In Proceedings of the 2013 IEEE Intelligent Vehicles Symposium (IV), Gold Coast, Australia, 23–26 June 2013; pp. 809–816.
4. Chan, C.Y. Advancements, prospects, and impacts of automated driving systems. *Int. J. Transp. Sci. Technol.* **2017**, *6*, 208–216. [CrossRef]

5. Singh, S. Critical Reasons for Crashes Investigated in the National Motor Vehicle Crash Causation Survey. In *Technical Report, National Highway Traffic Safety Administration*; 2015. Available online: https://crashstats.nhtsa.dot.gov/Api/Public/ViewPublication/812506 (accessed on 5 April 2019).

6. Zhang, F.; Clarke, D.; Knoll, A. Vehicle detection based on lidar and camera fusion. In Proceedings of the 17th International IEEE Conference on Intelligent Transportation Systems (ITSC), Qingdao, China, 8–11 October 2014; pp. 1620–1625.

7. Okumura, B.; James, M.R.; Kanzawa, Y.; Derry, M.; Sakai, K.; Nishi, T.; Prokhorov, D. Challenges in perception and decision making for intelligent automotive vehicles: A case study. *IEEE Trans. Intell. Veh.* **2016**, *1*, 20–32. [CrossRef]

8. Rao, A.S.; Georgeff, M.P. Modeling rational agents within a BDI-architecture. *KR* **1991**, *91*, 473–484.

9. Veres, S.M.; Molnar, L.; Lincoln, N.K.; Morice, C.P. Autonomous vehicle control systems—A review of decision making. *J. Syst. Control Eng.* **2011**, *225*, 155–195. [CrossRef]

10. Izzo, P.; Qu, H.; Veres, S.M. A stochastically verifiable autonomous control architecture with reasoning. In Proceedings of the 55th IEEE Conference on Decision and Control, CDC'16, Las Vegas, NV, USA, 12–14 December 2016; pp. 4985–4991.

11. Dennis, L.A.; Fisher, M.; Lincoln, N.K.; Lisitsa, A.; Veres, S.M. Practical verification of decision-making in agent-based autonomous systems. *Autom. Softw. Eng.* **2016**, *23*, 305–359. [CrossRef]

12. Seshia, S.A.; Sadigh, D.; Sastry, S.S. Formal methods for semi-autonomous driving. In Proceedings of the 2015 52nd ACM/EDAC/IEEE Design Automation Conference (DAC), San Francisco, CA, USA, 8–12 June 2015; pp. 1–5.

13. Luckcuck, M.; Farrell, M.; Dennis, L.A.; Dixon, C.; Fisher, M. Formal specification and verification of autonomous robotic systems: A survey. *ACM Comput. Surv. CSUR* **2019**, *52*, 1–41. [CrossRef]

14. SAE On-Road Automated Vehicle Standards Committee. Taxonomy and definitions for terms related to on-road motor vehicle automated driving systems. *SAE Stand. J.* **2014**, *3016*, 1–16.

15. Huang, X.; Kwiatkowska, M.Z. Reasoning about cognitive trust in stochastic multiagent systems. In Proceedings of the Thirty-First AAAI Conference on Artificial Intelligence, San Francisco, CA, USA, 4–9 February 2017.

16. Wooldridge, M. *An Introduction to MultiAgent Systems*; Wiley: Chichester, UK, 2002.

17. Lincoln, N.K.; Veres, S.M. Natural Language Programming of Complex Robotic BDI Agents. *J. Intell. Robot. Syst.* **2013**, *71*, 211–230. [CrossRef]

18. Dennis, L.A.; Aitken, J.M.; Collenette, J.; Cucco, E.; Kamali, M.; McAree, O.; Shaukat, A.; Atkinson, K.; Gao, Y.; Veres, S.M.; et al. Agent-based autonomous systems and abstraction engines: Theory meets practice. In *Annual Conference Towards Autonomous Robotic Systems*; Springer: Berlin/Heidelberg, Germany, 2016; pp. 75–86.

19. Kamali, M.; Dennis, L.A.; McAree, O.; Fisher, M.; Veres, S.M. Formal verification of autonomous vehicle platooning. *Sci. Comput. Program.* **2017**, *148*, 88–106. [CrossRef]

20. Al-Shihabi, T.; Mourant, R.R. A framework for modeling human-like driving behaviors for autonomous vehicles in driving simulators. In Proceedings of the Fifth International Conference on Autonomous Agents, Montreal, QC, Canada, 28 May–1 June 2001; pp. 286–291.

21. da Costa Sousa, J.M.; Palm, R.; Silva, C.; Runkler, T.A. Optimizing logistic processes using a fuzzy decision making approach. *IEEE Trans. Syst. Man Cybern. Part A Syst. Hum.* **2003**, *33*, 245–256. [CrossRef]

22. Palm, R.; Bouguerra, A.; Abdullah, M.; Lilienthal, A.J. Navigation in human-robot and robot-robot interaction using optimization methods. In Proceedings of the 2016 IEEE International Conference on Systems, Man, and Cybernetics (SMC), Budapest, Hungary, 9–12 October 2016; pp. 004489–004494.

23. Quigley, M.; Conley, K.; Gerkey, B.; Faust, J.; Foote, T.; Leibs, J.; Wheeler, R.; Ng, A.Y. ROS: An open-source Robot Operating System. In *ICRA Workshop on Open Source Software*; McGill University, Centre for Intelligent Machines (CIM): Kobe, Japan, 2009; Volume 3, p. 5.

24. Koenig, N.; Howard, A. Design and use paradigms for gazebo, an open-source multi-robot simulator. In Proceedings of the Intelligent Robots and Systems, (IROS 2004), Sendai, Japan, 28 September–2 October 2004; Volume 3, pp. 2149–2154.

25. Lomuscio, A.; Qu, H.; Raimondi, F. MCMAS: A model checker for the verification of multi-agent systems. In Proceedings of the International Conference on Computer Aided Verification, Grenoble, France, 26 June–2 July 2009; Springer: Berlin/Heidelberg, Germany, 2009; pp. 682–688.

26. Izzo, P.; Qu, H.; Veres, S.M. Reducing complexity of autonomous control agents for verifiability. *arXiv* **2016**, arXiv:1603.01202.

27. Kwiatkowska, M.; Norman, G.; Parker, D. PRISM 4.0: Verification of probabilistic real-time systems. In *Computer Aided Verification*; Springer: Berlin/Heidelberg, Germany, 2011; pp. 585–591.

28. Sadigh, D.; Sastry, S.; Seshia, S.A.; Dragan, A.D. Planning for autonomous cars that leverage effects on human actions. In *Robotics: Science and Systems*; University of Michigan: Ann Arbor, MI, USA, 2016; Volume 2.

29. McAree, O.; Aitken, J.M.; Veres, S.M. Towards artificial situation awareness by autonomous vehicles. *IFAC-Pap.* **2017**, *50*, 7038–7043.

30. Gleirscher, M.; Kugele, S. Defining risk states in autonomous road vehicles. In Proceedings of the 2017 IEEE 18th International Symposium on High Assurance Systems Engineering (HASE), Singapore, 12–14 January 2017; pp. 112–115.

31. Sadigh, D.; Landolfi, N.; Sastry, S.S.; Seshia, S.A.; Dragan, A.D. Planning for cars that coordinate with people: Leveraging effects on human actions for planning and active information gathering over human internal state. *Auton. Robot.* **2018**, *42*, 1405–1426. [CrossRef]

32. Li, L.; Ota, K.; Dong, M. Humanlike driving: Empirical decision-making system for autonomous vehicles. *IEEE Trans. Veh. Technol.* **2018**, *67*, 6814–6823. [CrossRef]
33. Al-Nuaimi, M.; Qu, H.; Veres, S.M. Computational Framework for Verifiable Decisions of Self-Driving Vehicles. In Proceedings of the 2018 IEEE Conference on Control Technology and Applications (CCTA), Copenhagen, Denmark, 21–24 August 2018; pp. 638–645.
34. Al-Nuaimi, M.; Qu, H.; Veres, S.M. A stochastically verifiable decision making framework for autonomous ground vehicles. In Proceedings of the 2018 IEEE International Conference on Intelligence and Safety for Robotics (ISR), Shenyang, China, 24–27 August 2018; pp. 26–33.
35. Levinson, J.; Askeland, J.; Becker, J.; Dolson, J.; Held, D.; Kammel, S.; Kolter, J.Z.; Langer, D.; Pink, O.; Pratt, V.; et al. Towards fully autonomous driving: Systems and algorithms. In Proceedings of the 2011 IEEE Intelligent Vehicles Symposium (IV), Baden-Baden, Germany, 5–9 June 2011; pp. 163–168.
36. Leonard, J.; How, J.; Teller, S.; Berger, M.; Campbell, S.; Fiore, G.; Fletcher, L.; Frazzoli, E.; Huang, A.; Karaman, S.; et al. A perception-driven autonomous urban vehicle. *J. Field Robot.* **2008**, *25*, 727–774. [CrossRef]
37. Urmson, C.; Anhalt, J.; Bagnell, D.; Baker, C.; Bittner, R.; Clark, M.; Dolan, J.; Duggins, D.; Galatali, T.; Geyer, C.; et al. Autonomous driving in urban environments: Boss and the urban challenge. *J. Field Robot.* **2008**, *25*, 425–466. [CrossRef]
38. Bhadani, R.K.; Sprinkle, J.; Bunting, M. The cat vehicle testbed: A simulator with hardware in the loop for autonomous vehicle applications. *arXiv* **2018**, arXiv:1804.04347.
39. Dosovitskiy, A.; Ros, G.; Codevilla, F.; Lopez, A.; Koltun, V. CARLA: An open urban driving simulator. In *Conference on Robot Learning, Proceedings of the Machine Learning Research*; 2017; pp. 1–16. Available online: http://proceedings.mlr.press/ (accessed on 5 April 2019).
40. Rong, G.; Shin, B.H.; Tabatabaee, H.; Lu, Q.; Lemke, S.; Možeiko, M.; Boise, E.; Uhm, G.; Gerow, M.; Mehta, S.; et al. Lgsvl simulator: A high fidelity simulator for autonomous driving. In Proceedings of the 2020 IEEE 23rd International Conference on Intelligent Transportation Systems (ITSC), Rhodes, Greece, 20–23 September 2020; pp. 1–6.
41. Chavez-Garcia, R.O.; Aycard, O. Multiple sensor fusion and classification for moving object detection and tracking. *IEEE Trans. Intell. Transp. Syst.* **2015**, *17*, 525–534. [CrossRef]
42. Park, W.J.; Kim, B.S.; Seo, D.E.; Kim, D.S.; Lee, K.H. Parking space detection using ultrasonic sensor in parking assistance system. In Proceedings of the 2008 IEEE Intelligent Vehicles Symposium, Eindhoven, The Netherlands, 4–6 June 2008; pp. 1039–1044.
43. Agarwal, V.; Murali, N.V.; Chandramouli, C. A cost-effective ultrasonic sensor-based driver-assistance system for congested traffic conditions. *IEEE Trans. Intell. Transp. Syst.* **2009**, *10*, 486–498. [CrossRef]
44. Kianpisheh, A.; Mustaffa, N.; Limtrairut, P.; Keikhosrokiani, P. Smart parking system (SPS) architecture using ultrasonic detector. *Int. J. Softw. Eng. Appl.* **2012**, *6*, 55–58.
45. Lee, M.; Kim, S.; Lim, W.; Sunwoo, M. Probabilistic occupancy filter for parking slot marker detection in an autonomous parking system using avm. *IEEE Trans. Intell. Transp. Syst.* **2018**, *20*, 2389–2394. [CrossRef]
46. Wikipedia. Self-Driving Car Incidents. 2019. Available online: https://en.wikipedia.org/wiki/List_of_self-driving_car_fatalities (accessed on 13 February 2021).
47. Fernandes, L.E.; Custodio, V.; Alves, G.V.; Fisher, M. A rational agent controlling an autonomous vehicle: Implementation and formal verification. *arXiv* **2017**, arXiv:1709.02557.
48. Giaquinta, R.; Hoffmann, R.; Ireland, M.; Miller, A.; Norman, G. Strategy synthesis for autonomous agents using PRISM. In *NASA Formal Methods Symposium*; Springer: Berlin/Heidelberg, Germany, 2018; pp. 220–236.
49. Trimble Inc. SkechUp Software. 2017. Available online: https://www.sketchup.com/ (accessed on 24 April 2018).
50. Qu, H.; Veres, S.M. Verification of logical consistency in robotic reasoning. *Robot. Auton. Syst.* **2016**, *83*, 44–56. [CrossRef]
51. Kwiatkowska, M.; Norman, G.; Parker, D. A framework for verification of software with time and probabilities. In *International Conference on Formal Modeling and Analysis of Timed Systems*; Springer: Berlin/Heidelberg, Germany, 2010; pp. 25–45.
52. Veres, S.M. *Natural Language Programming of Agents and Robotic Devices*; Sysbrain Ltd.: London, UK, 2008.
53. Seenouvong, N.; Watchareeruetai, U.; Nuthong, C.; Khongsomboon, K.; Ohnishi, N. A computer vision based vehicle detection and counting system. In Proceedings of the Knowledge and Smart Technology (KST), Chiangmai, Thailand, 3–6 February 2016; pp. 224–227.
54. Kato, T.; Guo, C.; Kidono, K.; Kojima, Y.; Naito, T. SpaFIND: An effective and low-cost feature descriptor for pedestrian protection systems in economy cars. *IEEE Trans. Intell. Veh.* **2017**, *2*, 123–132. [CrossRef]
55. Redmon, J.; Farhadi, A. YOLOv3: An Incremental Improvement. *arXiv* **2018**, arXiv:1804.02767.
56. Dollár, P.; Appel, R.; Belongie, S.; Perona, P. Fast feature pyramids for object detection. *IEEE Trans. Pattern Anal. Mach. Intell.* **2014**, *36*, 1532–1545. [CrossRef]
57. Zhang, J.; Singh, S. LOAM: Lidar Odometry and Mapping in Real-time. In *Robotics: Science and Systems*; University of California: Berkeley, CA, USA, 2014; Volume 2, p. 9.
58. Schwarting, W.; Alonso-Mora, J.; Rus, D. Planning and decision-making for autonomous vehicles. *Annu. Rev. Control. Robot. Auton. Syst.* **2018**, *1*, 187–210. [CrossRef]
59. Abbeel, P.; Ng, A.Y. Apprenticeship learning via inverse reinforcement learning. In Proceedings of the Twenty-First International Conference on Machine Learning, Banff, AB, Canada, 4–8 July 2004; ACM: New York, NY, USA, 2004; p. 1.

60. Karasev, V.; Ayvaci, A.; Heisele, B.; Soatto, S. Intent-aware long-term prediction of pedestrian motion. In Proceedings of the 2016 IEEE International Conference on Robotics and Automation (ICRA), Stockholm, Sweden, 16–21 May 2016; pp. 2543–2549.
61. Abbeel, P.; Dolgov, D.; Ng, A.Y.; Thrun, S. Apprenticeship learning for motion planning with application to parking lot navigation. In Proceedings of the 2008 IEEE/RSJ International Conference on Intelligent Robots and Systems, Nice, France, 22–26 September 2008; pp. 1083–1090.
62. Hubmann, C.; Schulz, J.; Becker, M.; Althoff, D.; Stiller, C. Automated driving in uncertain environments: Planning with interaction and uncertain maneuver prediction. *IEEE Trans. Intell. Veh.* **2018**, *3*, 5–17. [CrossRef]
63. Gindele, T.; Brechtel, S.; Dillmann, R. Learning driver behavior models from traffic observations for decision making and planning. *IEEE Intell. Transp. Syst. Mag.* **2015**, *7*, 69–79. [CrossRef]
64. Dräger, K.; Kwiatkowska, M.; Parker, D.; Qu, H. Local abstraction refinement for probabilistic timed programs. *Theor. Comput. Sci.* **2014**, *538*, 37–53. [CrossRef]
65. Paden, B.; Čáp, M.; Yong, S.Z.; Yershov, D.; Frazzoli, E. A survey of motion planning and control techniques for self-driving urban vehicles. *IEEE Trans. Intell. Veh.* **2016**, *1*, 33–55. [CrossRef]
66. Skiena, S. Dijkstra's algorithm. In *Implementing Discrete Mathematics: Combinatorics and Graph Theory with Mathematica*; Addison-Wesley: Reading, MA, USA, 1990; pp. 225–227.
67. Jihoonl. Dijkstra algorithm-ROS Package. 2014. Available online: http://wiki.ros.org/asr_navfn (accessed on 13 February 2018).
68. Rösmann, C.; Feiten, W.; Wösch, T.; Hoffmann, F.; Bertram, T. Efficient trajectory optimization using a sparse model. In Proceedings of the 2013 European Conference on Mobile Robots, Barcelona, Spain, 25–27 September 2013; pp. 138–143.
69. Roesmann, C. teb_local_planner. 2018. Available online: http://wiki.ros.org/teb_local_planner (accessed on 15 February 2019).
70. Adouane, L. *Autonomous Vehicle Navigation: From Behavioral to Hybrid Multi-Controller Architectures*; CRC Press: Boca Raton, FL, USA, 2016.
71. Rao, A.S.; Georgeff, M.P. An Abstract Architecture for Rational Agents. In *Principles of Knowledge Representation and Reasoning*; Morgan Kaufmann Publishers Inc.: Cambridge, MA, USA, 1992; pp. 439–449.
72. Bhat, A.; Aoki, S.; Rajkumar, R. Tools and methodologies for autonomous driving systems. *Proc. IEEE* **2018**, *106*, 1700–1716. [CrossRef]
73. Fisher, M.; Dennis, L.A.; Webster, M. Verifying Autonomous Systems. *ACM Commun.* **2013**, *56*, 84–93. [CrossRef]
74. Aitken, J.M.; Veres, S.M.; Shaukat, A.; Gao, Y.; Cucco, E.; Dennis, L.A.; Fisher, M.; Kuo, J.A.; Robinson, T.; Mort, P.E. Autonomous nuclear waste management. *IEEE Intell. Syst.* **2018**, *33*, 47–55. [CrossRef]
75. Hazim, M.Y.; Qu, H.; Veres, S.M. Testing, Verification and Improvements of Timeliness in ROS processes. In Proceedings of the Conference Towards Autonomous Robotic Systems, Sheffield, UK, 26 June–1 July 2016; Springer: Berlin/Heidelberg, Germany, 2016; pp. 146–157.
76. Ridel, D.; Rehder, E.; Lauer, M.; Stiller, C.; Wolf, D. A literature review on the prediction of pedestrian behavior in urban scenarios. In Proceedings of the 2018 21st International Conference on Intelligent Transportation Systems (ITSC), Maui, HI, USA, 4–7 November 2018; pp. 3105–3112.
77. Rasouli, A.; Kotseruba, I.; Tsotsos, J.K. Understanding pedestrian behavior in complex traffic scenes. *IEEE Trans. Intell. Veh.* **2017**, *3*, 61–70. [CrossRef]
78. Kooij, J.F.P.; Schneider, N.; Flohr, F.; Gavrila, D.M. Context-based pedestrian path prediction. In *European Conference on Computer Vision*; Springer: Berlin/Heidelberg, Germany, 2014; pp. 618–633.
79. Clarke, E.M.; Emerson, E.A.; Sistla, A.P. Automatic Verification of Finite-State Concurrent Systems Using Temporal Logic Specifications. *ACM Trans. Program. Lang. Syst.* **1986**, *8*, 244–263. [CrossRef]
80. Rangesh, A.; Trivedi, M.M. When Vehicles See Pedestrians with Phones: A Multicue Framework for Recognizing Phone-Based Activities of Pedestrians. *IEEE Trans. Intell. Veh.* **2018**, *3*, 218–227. [CrossRef]
81. Rasouli, A.; Kotseruba, I.; Tsotsos, J.K. Agreeing to cross: How drivers and pedestrians communicate. In Proceedings of the Intelligent Vehicles Symposium (IV), Los Angeles, CA, USA, 11–14 June 2017; pp. 264–269.
82. Saleh, K.; Hossny, M.; Nahavandi, S. Intent prediction of pedestrians via motion trajectories using stacked recurrent neural networks. *IEEE Trans. Intell. Veh.* **2018**, *3*, 414–424. [CrossRef]
83. Wiest, J.; Höffken, M.; Kreßel, U.; Dietmayer, K. Probabilistic trajectory prediction with Gaussian mixture models. In Proceedings of the 2012 IEEE Intelligent Vehicles Symposium, Alcala de Henares, Spain, 3–7 June 2012; pp. 141–146.

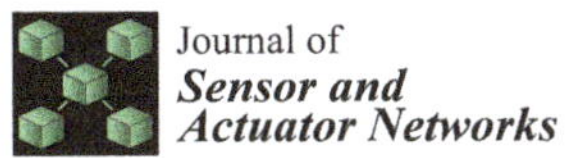

Journal of
Sensor and Actuator Networks

Article

A Double-Level Model Checking Approach for an Agent-Based Autonomous Vehicle and Road Junction Regulations

Gleifer Vaz Alves [1,*], Louise Dennis [2] and Michael Fisher [2]

[1] Graduate Program in Computer Science (PPGCC), Federal University of Technology—Parana (UTFPR), Ponta Grossa 84017-220, PR, Brazil

[2] Department of Computer Science, University of Manchester, Manchester M13 9PL, UK; louise.dennis@manchester.ac.uk (L.D.); michael.fisher@manchester.ac.uk (M.F.)

* Correspondence: gleifer@utfpr.edu.br or gleifervaz@gmail.com

Abstract: Usually, the design of an Autonomous Vehicle (AV) does not take into account traffic rules and so the adoption of these rules can bring some challenges, e.g., how to come up with a Digital Highway Code which captures the proper behaviour of an AV against the traffic rules and at the same time minimises changes to the existing Highway Code? Here, we formally model and implement three Road Junction rules (from the UK Highway Code). We use timed automata to model the system and the MCAPL (*Model Checking Agent Programming Language*) framework to implement an agent and its environment. We also assess the behaviour of our agent according to the Road Junction rules using a double-level Model Checking technique, i.e., UPPAAL at the design level and AJPF (*Agent Java PathFinder*) at the development level. We have formally verified 30 properties (18 with UPPAAL and 12 with AJPF), where these properties describe the agent's behaviour against the three Road Junction rules using a simulated traffic scenario, including artefacts like traffic signs and road users. In addition, our approach aims to extract the best from the double-level verification, i.e., using time constraints in UPPAAL timed automata to determine thresholds for the AVs actions and tracing the agent's behaviour by using MCAPL, in a way that one can tell when and how a given Road Junction rule was selected by the agent. This work provides a proof-of-concept for the formal verification of AV behaviour with respect to traffic rules.

Keywords: Rules of the Road; Autonomous Vehicles; agents; model checking

Citation: Alves, G.V.; Dennis, L.; Fisher, M. A Double-Level Model Checking Approach for an Agent-Based Autonomous Vehicle and Road Junction Regulations. *J. Sens. Actuator Netw.* **2021**, *10*, 41. https://doi.org/10.3390/jsan10030041

Academic Editors: Claudio Savaglio, Daniela Briola, Rafael C. Cardoso, Angelo Ferrando, Claudio Menghi and Tobias Ahlbrecht

Received: 15 March 2021
Accepted: 21 June 2021
Published: 25 June 2021

Publisher's Note: MDPI stays neutral with regard to jurisdictional claims in published maps and institutional affiliations.

1. Introduction

The deployment of Autonomous Vehicles (AVs) in urban road networks is possible in the near future. However, many challenges arise on the way towards this goal, for example: how can policy-makers ensure an AV is safe to operate within their jurisdiction [1]? This, and other complex issues, mean that the design and development of an AV must go through several stages involving a multistakeholder approach, which includes regulators, AV developers, safety experts, members of the public among others.

While the design of an AV should include sensors, cameras, software development, security protections, etc., it should also take into consideration the assessment of the traffic rules within which the AV will operate. If not, questions concerning safe operation cannot be properly answered. However, as highlighted by both Prakken [2] and Alves et al. [3,4] these traffic rules are rarely considered in the design and assessment of AVs.

1.1. Autonomous Vehicles and the Rules of the Road

Recent documents, such as [1,5,6] have started to highlight and discuss the need for a Digital Highway Code, where an AV would need to comply with the local traffic rules (or the "Rules of the Road"). It is well known that such a task brings challenges, mainly since Highway Codes were not designed to operate alongside autonomous sys-

J. Sens. Actuator Netw. **2021**, *10*, 41. https://doi.org/10.3390/jsan10030041

https://www.mdpi.com/journal/jsan

tems, but also since the description of the rules is predominantly human-readable, and not machine-readable.

So, how can we tackle such challenges? On the one hand, an AV should comply with traffic laws in a way that requires very few changes to create a Digital Highway Code [1]. On the other hand, it is understandable that new "Rules of the Road" may need to be designed in order to properly handle autonomous systems in urban traffic environments [6]. There is a clear trade-off between these two issues. Two examples illustrate the need to have "Rules of the Road" designed into AVs. As highlighted in ref. [1], in the state of Arizona (US) an AV operator may be issued a citation if the AV does not comply with traffic laws. A further example can be seen in the PAS-1882 standard from the BSI [7], which specifies "data collection and management for automated vehicle trials". In this standard several mechanisms are described for collecting the necessary data to conduct an AV trial. To the best of our knowledge, this standard currently does not contain data concerning questions such as when and how the "Rules of the Road" have been used by the AV and these could be quite important. These two examples remind us that, by ensuring the autonomous software abides by the "Rules of the Road" when AV is designed, it is definitely a useful asset for the stakeholders concerned with the proper behaviour of an AV on the roads.

In ref. [8], Waymo released a Safety Report on their vehicles. This document presents a reference from the *National Highway Traffic Safety Administration* (NHTSA), which shows the four scenarios that accounted for the vast majority of crashes in the US:

- 29% of the vehicles were involved in rear-end crashes;
- 24% of the vehicles were turning or crossing at intersections just before the crashes;
- 19% of the vehicles ran off the edge of the road;
- 12% involved vehicles changing lanes.

Consequently, Road Junction rules (which deal with crossing and entering intersections) provide a good case study for understanding the interaction of AVs and the Rules of the Road. This will enable us to develop an approach that can inform stakeholders around the development of Digital Highway Codes.

In our work, we select three Road Junction rules, from the UK Highway code [9], because road junction behaviour is a contributory factor in many crashes as discussed above [8]. We aim to embed these traffic rules into an agent, where this agent describes the basic behaviour of an AV in Road Junction scenarios. With this, we intend to determine: (i) Can these three selected road junction rules be used directly (i.e., as seen in the UK Highway code) by an AV? (ii) How to assess the AVs behaviour against the three road junction rules considering simple Road Junction scenarios? and (iii) Are there any guidelines that can be given to enable the AV to work correctly with such Road Junction rules?

1.2. Related Work

Considering the related work, there are Rizaldi et al. [10] and Bhuiyan et al. [11] that present a formalisation for road traffic rules. Nonetheless, neither approach uses an agent abstraction to represent the AV behaviour and decision-making. Besides, Kamali et al. [12] and Al-Nuaimi et al. [13] both present the use of agents to model the AV and the formal verification of agents. However, their AV application scenario is not related to the "Rules of the Road". So, our work aims to formalise the Road Junction rules, use an agent to represent an AV and apply formal verification techniques to properly assess an agent's behaviour in road traffic scenarios .

In ref. [14], Bakar and Selamat present a systematic literature review of agent systems verification, where they describe the most used techniques to formally verifying agents. Their figures show that 49% of techniques are applied at the design stage, 27% during development, and 25% at runtime. Model Checking or model-based verification techniques are used in 44% of the work, while most of the properties verified are temporal ones (19%), followed by epistemic properties (9%). These figures serve to endorse our choice of a double-level Model Checking, i.e., applying formal verification of temporal properties at both design and development levels.

1.2.1. Proposal

Figure 1 shows our proposed SAE-RoR (*Simulated Automotive Environment for the Rules Of the Road*) architecture. In previous work [4], we presented the first version of SAE-RoR architecture, where we focused on the formalisation of the Road Junction rules using Linear Temporal Logic (LTL) and also the first steps towards the implementation of a single rule using the agent programming language, GWENDOLEN [15].Now, we extend the SAE-RoR architecture by adding an extra layer of modelling with timed automata and Model Checking using UPPAAL [16]. We have also added two further Road Junction rules and the formal verification of properties using AJPF, which is responsible for Program Model Checking of the GWENDOLEN implementation [17].

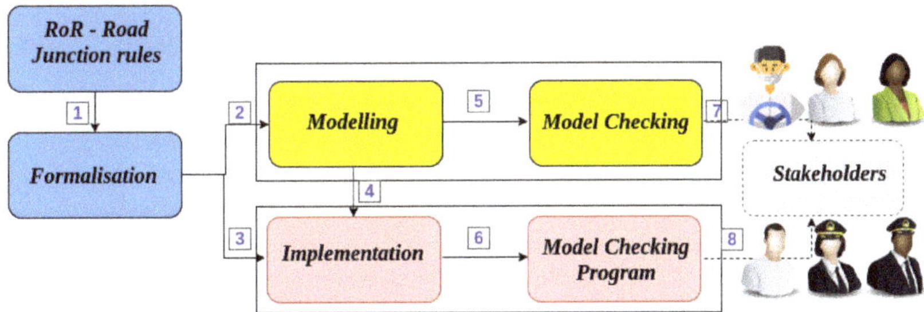

Figure 1. Proposal: SAE-RoR architecture.

Each element of the SAE-RoR architecture is described in detail in Sections 4–6. Here, we explain the general workflow using the architecture in Figure 1. Step 1, i.e., arrow 1 between the two blue components in Figure 1, represents the formalisation using LTL of the Road Junction rules from the UK Highway Code (this was initially described in [4]). This LTL formalisation helps us to abstract the informal concepts and elements in the UK Highway Code to an unambiguous formal language. Then steps 2 and 3 represent the use of this language as a basis to respectively build the UPPAAL model and the GWENDOLEN implementation. Notice that step 4 shows the mapping from UPPAAL model elements to the agent's implementation components. Next, step 5 represents the formal verification of properties of the model using the UPPAAL model checker, while step 6 shows those properties concerned with the agent's implementation that are verified using the AJPF model checker. Steps 7 and 8 describe stages of the SAE-RoR architecture that are not implemented yet (forming future work). Nonetheless, this Stakeholders stage is an important feature of our proposal. Steps 7 and 8 represent the outcomes from property verification that could guide the actions of a given Stakeholder. As an example, a Policymaker could use a counter-example describing a safety violation from a given model and decide whether a traffic rule concerning pedestrians needs to be changed. In ref. [6], some possible Stakeholders (related to AVs) are mentioned: Driver, Road User, Safety Expert, AV developer, Researcher, Policymaker, Traffic Officer, Emergency services and police, Local government, Highway authorities, Public sector, Insurance, Politicians, Legal, among others. These stakeholders form suitable end-users for our proposed workflow.

1.2.2. Contributions

The main contributions of our work are the following:

1. A complete architecture for the Formal Verification of Agents in the Rules of the Road, which starts at the formalisation of Road Junction rules, passes these to modelling and implementation tools, generates formal verification results, which may be of interest to the given stakeholders.

2. The double-level Model Checking approach, which makes it possible to formally verify properties both at design and development levels. As proof-of-concept, we have verified 30 properties, 18 at design level and 12 at development level.

3. A set of verified properties, where properties range from time constraints to the analysis of the AV-agent's behaviour considering all possible actions the agent can take in Road Junction scenarios.

4. The creation of an agent's environment that includes random generation of events (following the methodology outlined in ref. [18]), where different scenarios concerning selected Road Junction rules are simulated.

5. The use of a mapping from a given UPPAAL timed automaton to the basic elements of a BDI (Belief-Desire-Intention) agent.

6. Implementation of a BDI agent (in GWENDOLEN), which enables tracing of an agent's behaviour. Taken with the model-checking process (via AJPF) it is possible to assess what were the choices selected (autonomously) by the agent that led to any given outcome [19]. For example, given a certain scenario with specific perceptions from the environment, did the agent (correctly) choose to follow a given road junction rule?

1.2.3. Remarks and Limitations

It is necessary to determine some remarks and assumptions about our proposal, making it clear what is included (or not) in our model and implementation. Further, some limitations of our work are mentioned below.

- Single Agent: we only model a single agent in our implementation. And this agent has no (internal) concurrency to try to enter the Road Junction.
- No driving behaviour: this single agent has no driving behaviour component, the agent is only concerned with obeying the programmed Road Junction rules. So, we do not verify, for example, the speed or trajectory of the vehicle.
- No collision-freedom: we do not model collision avoidance behaviour here, though we have considered it in previous work [20,21].
- Intersection Management: we are not trying to deal with the well-known problem of Intersection Management using multiple agents [22]. We consider only the behaviour of a single agent following the desired traffic rules.
- Road Junction environment: The environment is represented in both modelling and simulation as a simple (9×9) grid with a few road features such as `stop signs`, and other road users. This is because this captures the abstract issues and, on a practical level, model-checking does not scale well once the grid size increases dramatically. Our environment model uses as basis the formalised abstract model which captures the temporal elements from the road junction rules. So, we do not represent spatial elements, as seen, for example in [23], where the author uses the Multi-lane Spatial Logic to represent virtual lanes in an urban traffic environment.
- Time constraints: we use time constraints in our model to represent thresholds within or after which actions and events should occur. These constraints are only used in our model, they are not taken from any Highway Code. This illustrates the way modellers need to extract implicit assumptions from rules written for human consumption—in this case that actions such as waiting will take a reasonable amount of time, or will take place within a reasonable space of time.
- Subset of Road Junction Rules: the UK Highway code has a set of 14 rules for Road Junctions. Here, we model and implement only the first three rules from the Road Junction set, specifically rules 170, 171, and 172.
- Modularity of the model checking stages: our two Model Checking stages (using UPPAAL and MCAPL) are loosely coupled, i.e., they take place independently of each other. On the one hand, this independence loses the potential benefit an integrated model checking semantics, where one could verify the whole system (model plus implementation). On the other hand, this modular architecture allows a separation of

concerns meaning a user can consider separately the verification of either the timed automata model verification or the agent's implementation.

- Nature of verified properties: all the verified properties are related to the basic behaviour of our agent against the three selected road junction rules in our simple road junction environment.

In our previous publications [3,4], we have presented the formalisation of the Road Junction rules in LTL and a partial implementation of Rule 170 in the Gwendolen language. Thus, this paper extends these initial results by the following:

- Including two new rules (171 and 172) from the UK Highway Code.
- Modelling the AV-agent and Road Junction environment using timed automata.
- Formal Verification of 18 properties of these timed automata using UPPAAL.
- Implementation of all three rules in Gwendolen.
- Formal Verification of 12 properties of this agent implementation using AJPF.

The remainder of this paper presents, in Sections 2 and 3, some useful terminology and background information. Next, Sections 4–6 describe the main stages of our work on modelling, implementation and formal verification results. In Section 7, related work is discussed while Section 8 provides final remarks.

2. Key Terminology & the Rules of the Road

In this section, we show the Road Junction rules that are used in our work. Before proceeding, we clarify some terms applied here to guide the description of our modelling, implementation, and verification.

2.1. Terminology and Abbreviation

- SAE-RoR: is the name of our proposed architecture (seen in Figure 1).
- AV: According to Herrmann et al. [24], the term automated vehicles refers to autonomy levels 1–4, while the terms autonomous, self-driving or driverless vehicles refer to autonomy level 5. Here, for the sake of simplicity, we only use the term Autonomous Vehicles (AV). And in our model, we are not concerned whether our agent represents a vehicle with level 4 or 5, or if the vehicle has a human driver responsible or only passengers inside it.
- AV-agent: is the name of our agent implemented in Gwendolen.
- AV_agent: is the name of the automaton modelled with UPPAAL, which represents the AV-agent.
- RU: according to the UK Highway Code, a Road User (RU) can be any of the following: pedestrian, cyclist, motorcyclist, powered wheelchair/mobility scooter, horse rider, etc.
- ru: control variable that represents a Road User and it is used in the UPPAAL timed automaton model.
- RJ: here we use the term Road Junction (RJ) which has the same meaning as an Intersection.
- RoR: we use the term "Rules of the Road" (RoR) which has the same meaning as traffic rules, Highway Code, traffic laws, or road traffic laws.
- Digital Highway Code: is the version of the Rules of the Road which is intended to work for AVs.
- Cross junction: a crossroad is the place where two roads meet and cross each other. It could be in the form of: a major road crossing a minor road; or two equal roads crossing each other [25].
- T junction: is a place where two roads meet in the shape of letter T [25].

2.2. The Road Junction Rules

The UK Highway Code has different sections, concerning Overtaking, Roundabouts, Road Junctions, among others [9]. We are focused on the Road Junction rules, which has 14 rules, from 170 to 183, describing when and how a driver is supposed to enter a road junction, to turn right, to turn left, to enter a road junction with traffic lights, among other situations. Here, we describe the three simple Road Junction rules that are modelled, implemented, and verified, rules: 170, 171 and 172. Before presenting the formalisation of these three rules we describe the LTL (Linear Temporal Logic) [26] operators and constants that we use in our formalisation for the Road Junction rules. Further details about this formalisation can be found in Alves et al. [4].

- Propositional operators from LTL:

  $\wedge, \vee, \rightarrow, \neg$.

  where these four propositional logical operations represent conjunction, disjunction, implication, and negation.

  $\Box, \Diamond, \bigcirc, \cup$.

  where these four future-time operators represent: always, eventually, next, and until.

2.2.1. LTL Formalisation

- Rule 170—UK Highway Code:

  – You should watch out for road users (RU).
  – Watch out for pedestrians crossing a road junction (JC) into which you are turning. If they have started to cross they have priority, so give way.
  – Look all around before emerging (NB: For the sake of clarity, we choose to use the term `enter` as an action which represents not only a driver entering a road junction, but also emerging from a road junction to another road). Do not cross or join a road until there is a safe gap (SG) large enough for you to do so safely.

- Rule 170, represented in LTL, describes when the autonomous vehicle (AV) may enter the junction (JC):

$$\Box\,((\texttt{watch(AV, JC, RU)} \wedge (\neg\,\texttt{cross(RU, JC)} \wedge (\texttt{exists(SG, JC)})) \rightarrow ((\texttt{exists(SG, JC)} \wedge \neg\,\texttt{cross(RU, JC)}) \cup \texttt{enter(AV, JC)})))$$

  `Informal Description:` it is always the case that the AV is supposed to watch for any road users (RU) at the junction (JC) and there are no road users crossing the junction and there is a safe gap (SG). Then, no road users crossing the junction and the existence of a safe gap should remain true, until the AV may enter the junction.

- Rule 170 represented in LTL, when the autonomous vehicle (AV) should give way at the junction (JC):

$$\Box\,(\texttt{watch(AV,JC,RU)} \wedge (\texttt{cross(RU,JC)}) \rightarrow \texttt{give-way(AV,JC)})$$

  `Informal Description:` it is always necessary to watch out for road users (RU) and check if there is a road user crossing the junction. Then, the AV should give way to traffic.

- Rule 171—UK Highway Code:

  – You MUST stop behind the line at a junction with a 'Stop' sign (ST) and a solid white line across the road. Wait for a safe gap (SG) in the traffic before you move off.

- Rule 171 represented in LTL:

$$\texttt{exists(ST,JC)} \rightarrow \Box\,(\texttt{stop(AV,JC)} \cup (\texttt{exists(SG,JC)} \wedge (\texttt{exists(SG,JC)} \cup \texttt{enter(AV,JC)})))$$

> Informal Description: when there is a stop sign (ST), then it is always the case the AV should stop at the junction until there is a safe gap (SG). And the safe gap must remain true until the AV enter at the junction.

- Rule 172—UK Highway Code:

  - The approach to a junction may have a 'Give Way' sign (GW) or a triangle marked on the road (RO). You MUST give way to traffic on the main road (MR) when emerging from a junction with broken white lines (BWL)across the road.

- Rule 172 represented in LTL:

$$\Box \, ((\texttt{exists(AV,RO)} \wedge \texttt{enter(AV,JC)})$$
$$\wedge \, ((\texttt{exists(BWL,JC)} \vee \texttt{exists(GW,JC)}) \rightarrow \texttt{give-way(AV,MR)}))$$

> Informal Description: It is always the case that when there is an AV driving on a Road (RO) and the AV enters the junction. And there is a Broken White Line (BWL) or a Give Way sign (GW), then the AV should give way to the traffic on the Main Road (MR).

2.2.2. Remarks

The LTL formalisation aims to describe most of the elements from the rules as given in the UK Highway Code, but some level of abstraction is needed to properly determine the formalisation. And when we build the automata models in UPPAAL and also the Gwendolen implementation of the AV-agent we have abstracted some additional elements from the formalised Road Junction rules, in a way that the three rules (170, 171, and 172) have been wrapped to work together and represent the possible behaviour of an AV alongside the Road Junction rules. Notice that this degree of abstraction (used in our approach) does not avoid the proper verification of the agent's behaviour to tell which rules have been selected by the agent. An example of such abstraction is noted in rule 172, where there are two different terms Give Way sign and triangle marked on the road with the same meaning. So, we only use the former term in our model and implementation.

3. Background

In this section we present some notation and concepts related to the models, languages and tools used in this work.

3.1. Timed Automata, Temporal Logic and UPPAAL

As presented in Baier and Katoen [27], timed automata model the behaviour of time-critical systems. A timed automaton has a finite set of clock variables. All clocks proceed at rate one. The value of a clock denotes the amount of time that has elapsed since its last reset. Conditions which depend on clock values are called clock (or time) constraints.

Definition 1 (Clock constraint). *A clock constraint over a set C of clocks is formed according to the grammar g:*

$$g ::= x < c \mid x \le c \mid x > c \mid x \ge c \mid g \wedge g$$

where $c \in \mathbb{N}$ and $x \in C$. $CC(C)$ represents the set of clock constraints over C.

The Timed Automaton definition [27] is given below.

Definition 2 (Timed Automaton). *A timed automaton is a tuple $TA = (Loc, Act, C, \hookrightarrow, Loc_0, Inv, AP, L)$ where*

- *Loc is a finite set of locations,*
- *$Loc_0 \subseteq Loc$ is a set of initial locations,*
- *Act is a finite set of actions,*
- *C is a finite set of clocks,*

- $\hookrightarrow \subseteq Loc \times CC(C) \times Act \times 2^C \times Loc$ *is a transition relation,*
- $Inv{:}Loc \rightarrow CC(C)$ *is an invariant-assignment function,*
- *AP is a finite set of atomic propositions, and*
- $L{:}Loc \rightarrow 2^{AP}$ *is a labelling function for the locations.*

ACC(TA) denotes the set of atomic clock constraints that occur in either a guard or a location invariant of TA.

Baier and Katoen [27] mention that *Timed Computation Tree Logic* (TCTL) is a real-time variant of temporal logic used to express properties of timed automata. So, the UPPAAL Model Checker which makes use of timed automata also uses a simplified version of TCTL to specify verification properties. Below, the syntax of TCTL is given (as seen in [27]) and also the corresponding syntax used in UPPAAL is provided in Table 1.

Definition 3 (Timed CTL syntax). *Formulae in TCTL are either state or path formulae. TCTL state formulae over set AP of atomic propositions and set C of clocks are formed according to the Φ grammar:*

$$\Phi ::= true \mid a \mid g \mid \Phi \wedge \Phi \mid \neg\Phi \mid \exists\varphi \mid \forall\varphi$$

where $a \in AP$, $g \in ACC(C)$ and φ is a path formula defined by:

$$\varphi ::= \Phi \,\bigcup\nolimits^{J}\, \Phi$$

where $J \subseteq \mathbb{R}_{\geq 0}$ is an interval whose bounds are natural numbers.

NB: the propositional logic operators $(\vee, \rightarrow, etc)$ are obtained from $\vee$ and $\neg$. Also, the temporal logic operators $\square$ and $\Diamond$ are obtained by using existing operators in Φ and φ grammars.

In Table 1, we show the UPPAAL syntax (based on TCTL) used to write formulae and temporal properties.

Table 1. UPPAAL syntax.

Operator	Meaning
&&	And
||	Or
==	Equivalence
imply	Conditional
not	Negation
A	Universal quantifier
E	Existential quantifier
[]	Always
<>	Eventually
-->	Leads to

Below, we show an example of a formula written using UPPAAL syntax.

```
A[] (AV.at_roadjunction imply AV.enter_roadjunction)
```

This formula states: for all possible paths it is always the case that if the AV is placed at the road junction it will enter the road junction.

NB: The above example could be slightly changed to: $\forall\square$(`AV.at_roadjunction`$\rightarrow$ $\forall\Diamond$`AV.enter_roadjunction`), using TCTL notation. However, UPPAAL does not allow nesting of path formulae in a way that to write this formula, it is necessary to use the operator leads to ($\rightsquigarrow$). The previous TCTL formula may expressed using UPPAAL syntax as: `AV.at_roadjunction --> AV.enter_roadjunction`.

3.2. BDI Model and GWENDOLEN *Language*

In our work we use the GWENDOLEN agent programming language [15] in order to implement a BDI agent [28] to capture the core decision-making behaviour of an autonomous vehicle. By using GWENDOLEN, we can also take advantage of the MCAPL framework [17], where the AJPF model checker can be used to formally verify the behaviour of the agent. The MCAPL framework allows us to program BDI agents in languages such as GWENDOLEN and Goal [29], and one can also program the agent's environment using Java. In addition, it is possible to use the AJPF model checker to verify the agent's programming, where it is possible to check the agent's behaviour. AJPF is an extension of Java PathFinder (JPF) [30] which is, in turn, a tool for model-checking Java programs.

3.2.1. BDI Model

As described in Bordini et al. [31], the Beliefs-Desires-Intentions (BDI) Model is based on a model of human behaviour developed by Bratman [28].

- Beliefs are information the agent has about the world.
- Desires are all possible states of events that the agent might want to achieve.
- Intentions are the state of events the agent has decided to commit towards. These events can be goals that are assigned to the agent or the agent may choose among a set of options.

When implementing a BDI model in an agent programming language we usually have the following structure for an agent plan:

```
trigger_event : guard <- body
```

where a given agent may have different plans in order to achieve a certain goal.

- The `trigger_event` is given by a new belief or a goal.
- The `guard` is defined by a set of beliefs.
- The `body` is represented as a set of actions.

Example

We provide a simple example considering the AV-agent at a road junction.

```
AV-agent believes it is at the road junction.
AV-agent selects the intention to enter the road junction.
AV-agent triggers the following plan:
enter-roadjunction : at-roadjunction <- check-sign, watchout-for-road-user;
```

In this example initially the AV-agent believes it is placed at the road junction, next it has the desire to enter the road junction and selects an intention to achieve this goal. As a consequence it triggers a plan to execute two actions: the first one checks the existing traffic sign at the road junction and the second action is responsible for watching out for any road user crossing the junction.

3.2.2. GWENDOLEN Language

GWENDOLEN is an agent declarative logic-programming language incorporating explicit representations of goals, beliefs, and plans. The language uses similar syntactic conventions to other BDI agent languages. Here, we describe the syntax elements used in our implementation:

+b adds the belief b.
–b removes the belief b.
+!g adds the goal g.
+!g[perform] adds a new goal of type perform. Perform goals are discharged by the execution of an appropriate plan.

+!g[achieve] adds a new goal of type achieve. Achieve goals are discharged only when they become beliefs.

B x represents a guard condition, checks if belief x is perceivable.

G x represents a guard condition, checks if goal x has been added.

`hello(x)` represents that action `hello(x)` (defined in the agent's environment) is executed.

A plan in GWENDOLEN uses the syntax previously presented in a BDI model.

Example

We retake the previous example of the AV-agent, but now using GWENDOLEN syntax.

```
at(roadjunction)              \\ predefined belief ''agent is at the road junction''
enter-roadjunction[achieve] \\ a goal (of type achieve) to ''enter the road junction''
+!enter-roadjunction[achieve] : { B at(roadjunction) } <-
check-sign(A,B), watchout-for-road-user(C,D);
```

In the last lines (above) there is a GWENDOLEN plan that represents the following: when the agent recognises the trigger event (i.e., the achieve goal of entering the road junction), it checks the guard (i.e., the predefined belief which says the agent is at the road junction), and then the agent executes two actions: `check-sign(A,B)` and `watchout-for-road-user(C,D)`. These actions are implemented in the agent's environment. NB: the values A, B, C, D represent coordinates in a grid which represents a road junction environment.

3.3. The Property Specification Language

The MCAPL framework provides a Property Specification Language (PSL) used to write properties for the AJPF Model Checker. In Table 2, we present the set of operators from PSL which is used in the verification of properties.

Table 2. PSL operators.

Operator	Meaning
<>	temporal logic operator Eventually
[]	temporal logic operator Always
B	a Belief of the agent
G	a Goal of the agent
I	an Intention of the agent
D	an Action of the agent
ItD	an Intention to execute an action of the agent
P	a Perception from the environment
&	logical operator And
||	logical operator Or
-->	logical operator Implies

Example

Using the same elements from the two previous examples, we can write a PSL specification.

```
<>((B(AV-agent,at(roadjunction))) & D(AV-agent,watchout-for-roaduser(1,0)))
```

The description of this specification is: eventually the AV-agent believes it is at the road junction and the AV-agent executes the action `watchout-for-roaduser` at postion $(1, 0)$ (in the grid environment).

4. Modelling Using Timed Automata

The modelling of our system was carried out using timed automata within UPPAAL model checker tool. We have divided our model into two main automata: `AV_agent` and

RJ_Env (*Road Junction environment automaton*); and three additional (simple) automata which model specific artefacts from the road junction environment: road_user (*Watch out Road User automaton*), safe_gap (*Check for a Safe Gap automaton*), and sign (*Check traffic sign automaton*). These five automata set up a network of automata, which can all communicate with each other through synchronized channels. In our model, RJ_Env forms the main communication hub among all automata, receiving information from the artefacts as well as sending information to, and receiving information from, the AV_agent.

4.1. *AV_agent Automaton*

Figure 2 presents the automaton which models the basic behaviour of the AV-agent. The agent starts in a state where it is away from the RJ, the agent uses the communication channel to tell the RJ_Envautomaton that it is going to approach the RJ. Once the agent is at the RJ, it will receive from the RJ_Env one of two possible alternatives that may exist at the RJ: (i) there is only the stop sign; or (ii) there are both the stop and the give way signs. At this moment, the clock (x) starts to work and the agent is at the state of watching out for RU. From this state, there are two possible outcomes: (i) RJ_Env tells the agent that RJ is free; or (ii) RJ_Env tells the agent RJ is busy. When the latter occurs the agent is supposed to start to wait until it is possible to watch again for road users. When the RJ is free, the agent checks for a safe gap and again two outcomes are possible: (i) there is a safe gap and the RJ_Env tells the agent to enter the RJ; or (ii) there is no safe gap, the agent should wait, so it moves to the waiting state (the same one which is used when the RJ is busy).

After this, the AV-agent has successfully entered the RJ, and it tells the RJ_Env that it is now away from the RJ once more.

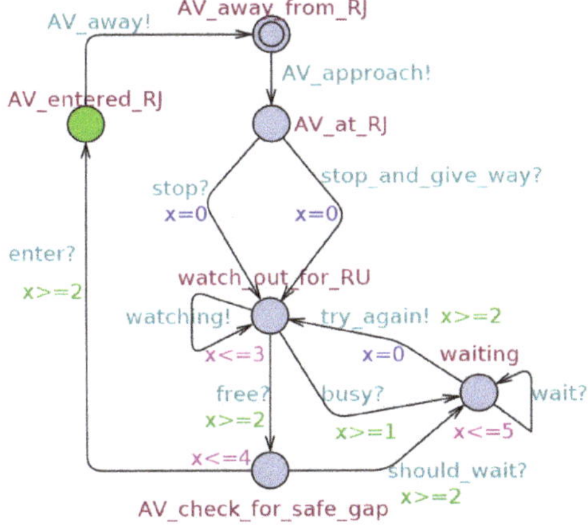

Figure 2. UPPAAL template for AV_agent automaton.

Time Constraints

To properly represent an AV, we have decided to add some time constraints to simulate thresholds for each one of the main actions in the system, i.e., watching for road users, waiting (at the RJ), check for a safe gap and entering the RJ.

Figure 3 illustrates how these time constraints work for the corresponding synchronized channels: busy, free, enter, should_wait, and try_again. The time constraints establish the lower and upper bounds for each one of the synchronized channels.

The lower and upper bounds for these time constraints have been selected considering that we could have a cross or a T junction, where the AV-agent should watch for road users at least in two different directions (in case of T junction) or at most in three directions (in case of a cross junction). However, in case the road junction (either cross or T junction) is busy, the AV-agent may only look once for road users and already check the RJ is busy.

Thus, the lower bound for the busy channel is 1 (one). For the remaining channels (enter, should_wait, and try_again) we only need to add one or two extra time units. The idea behind these extra time units is to model the additional time required for the AV-agent to execute its actions. For example, once the AV-agent checks the junction is free, then it will need an extra time unit to actually move and enter the junction.

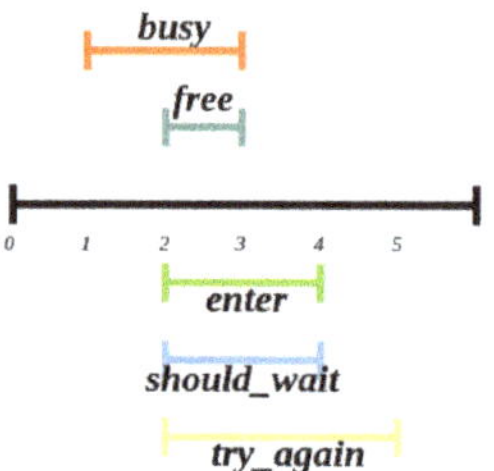

Figure 3. Time constraints for the `AV_agent` automaton synchronized channels.

4.2. Road Junction Environment Automaton

In Figure 4, the Road Junction environment automaton (`RJ_Env`) is shown. This model represents the behaviour and communication that the environment engages with the `AV_agent` and also with the three artefacts.

It starts in an idle state, as soon as the `AV_agent` approaches the RJ, the environment should check for the traffic sign, according to the information received from the `sign` artefact, the `RJ_Env` will tell the `AV_agent` if there is a single stop sign or two signs (stop and give way). After that, the `RJ_Env` waits for the `AV_agent` to start watching the RJ for road users. Now, two possible outcomes may be received from the `road_user` artefact: (i) the RU is away from the RJ; or (ii) the RU is crossing the RJ. When the latter occurs, the environment will notify the `AV_agent` that the RJ is busy, therefore the `AV_agent` is supposed to wait. Next, the `RJ_Env` waits to receive from the `AV_agent` the communication stating that it wants to try again and restart the checking for RU. But, in case the RJ is free, the environment will check for a safe gap with the corresponding `safe_gap` artefact. This artefact will answer whether or not there is a safe gap at the RJ. If there is no safe gap, the `RJ_Env` tells the `AV_agent` that it should wait at RJ. But, if there is a safe gap, thus the `RJ_Env` tells the `AV_agent` to enter the RJ. Finally, when the `AV_agent` tells the `RJ_Env` that it is away from the RJ, the environment is back to the idle state.

Notice that we use a variable ru (stands for Road User), which is incremented every time it perceives there is a Road User crossing the junction and is decreased every time there is a Road User away from the junction. So, before the synchronisation channel with the `AV_agent` is set up to communicate that the junction is free, it is necessary to check if ru is equal to zero (i.e., there is no Road User at all). If ru is not zero, there is still some Road User at the junction and the model does not proceed to the next stage, which is to check for a safe gap. NB: in the stage of verifying properties (see Section 6.1) we run simulations with one, two, and three Road Users, that is why we need this control variable ru.

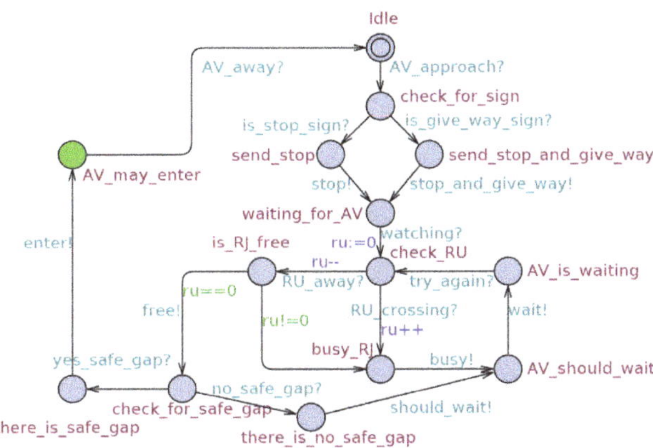

Figure 4. UPPAAL template for `RJ_Env` automaton.

4.3. Automata for the Artefacts

Figure 5 presents the three UPPAAL automata responsible for representing the behaviour of the artefacts from the environment (`RJ_Env`). The leftmost automaton shows the `sign` artefact, which should tell the environment the existing traffic sign (only stop sign or a stop and a give way signs). The centre automaton presents the `road_user` artefact, this one will send to `RJ_Env` the road user state (it is away from RJ or it is crossing). The rightmost automaton pictures the `safe_gap` artefact, which is responsible for telling whether or not there is a safe gap.

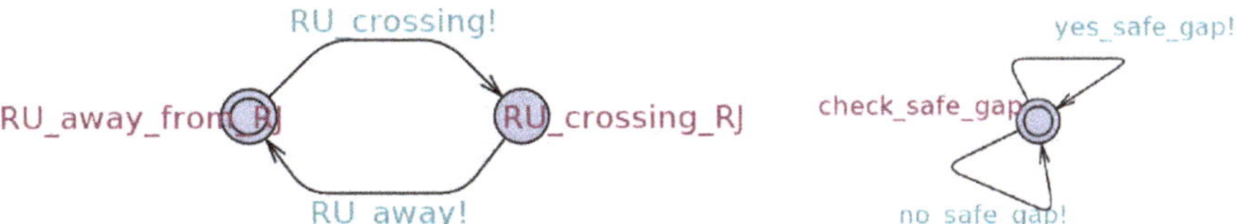

Figure 5. UPPAAL templates for the environment artefacts.

5. Agent and Environment Implementation

Our SAE-RoR system was implemented using the GWENDOLEN agent programming language and MCAPL framework. As previously shown in Figure 1, the implementation is based on LTL formalisation of RJ rules together with a mapping from the UPPAAL timed automata to the agent's implementation (this is presented later in Section 5.2). Nevertheless, some modifications were necessary because of the differences between UPPAAL and MCAPL frameworks. For example, by using UPPAAL we have modelled time constraints to represent the behaviour of the AV-agent in the road junction, while in MCAPL we have used random generation of events in the environment. In the following, we also describe the RJ environment modelling, implementation, and testing scenarios.

5.1. Setting-Up the Road Junction Environment Model

The model implemented using MCAPL is a simple representation of a crossroad junction. Figure 6 shows the grid which splits the road junction into nine spots. The grid set-up is as follows:

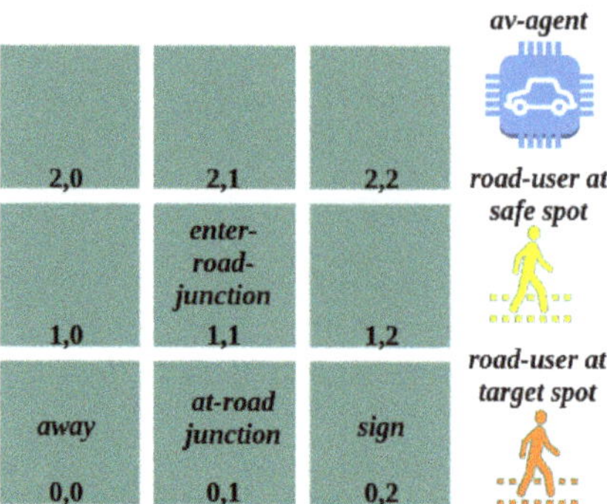

Figure 6. Road Junction grid environment model.

- spot (0,0): the start position for the AV-agent, when it is said to be away from the road junction.
- spot (0,1): the position the AV-agent goes when it is supposed to watch for road users.
- spot (0,2): the position where the traffic sign is placed.
- spot (1,1): the position reached by the AV-agent once it enters the RJ. Notice that after reaching (1,1) spot, the AV-agent may go to any of the following spots: {(1,0); (2,1); (1,2)}.
- spots {(1,0)); (1,1)); (2,1); (1,2)} are said to be target spots. That is, spots that can be reached by the AV-agent.
- spots {(0,0); (2,0); (2,2)} are said to be safe spots. That is, spots that can not be reached by the AV-agent, once it arrives at the road junction, i.e., AV-agent is placed at (0,1).
- a given road user may be placed at a safe or a target spot.

The above grid setup is implemented in the AV-agent as a set of initial beliefs.

5.2. Correspondence: Modelling and Implementation

Here we describe the correspondences from the UPPAAL modelling to the GWENDOLEN implementation. In Table 3, we describe the mapping from the AV_agent UPPAAL Automaton (previously seen in Figure 2) to the AV-agent implemented in GWENDOLEN (previously shown in Listing 1).

Notice that the information in the table is separated into *names* and *types* both for the UPPAAL model and the agent's implementation. For the model, the *names* represent the *Locations* or the *communication channels* used in the AV_agent Automaton, while the *types* identify which element this name represents, it can be a *Location* (**Loc**) or a communication channel with other UPPAAL Automata. In this case, we use the following representation:

UA1 is the AV_agent UPPAAL Automaton.
UA2 is the RJ_Env UPPAAL Automaton (see Figure 4).
UA3 is the sign UPPAAL Automaton (see Figure 5).
UA4 is the road_user UPPAAL Automaton (see Figure 5).
UA5 is the safe_gap UPPAAL Automaton (see Figure 5).

NB: when there are two pairs of different Automata as types, e.g., UA1-UA2/UA2-UA3 in the fourth row of the table. This means, the channel stop? is a communication from UA1 to UA2, while the channel is_stop_sign? synchronises the automata UA2 to UA3.

For the implementation, the *names* represent elements used in the GWENDOLEN code. These elements can be any of the following *types*:

Belief: represents an initial belief of the agent.

Add Belief: represents a new belief acquired by the agent during execution.
Percept: is a perception obtained by the agent from the environment.
Action: is an action executed by the agent in the environment.

Table 3. Correspondence between Model and Implementation.

Model		Implementation	
Names	**Types**	**Names**	**Types**
AV_away_from_RJ	Loc	av_away()	Belief
AV_approach!	UA1	approach_roadjunction()	Action
AV_at_RJ	Loc	at_roadjunction()	Percept
stop?/is_stop_sign?	UA1-UA2/U2-UA3	check_sign()/stop_sign()/stopped	Action/Percept/Add Belief
stop_and_give_way?/is_give_way_sign?	UA1-UA2/U2-UA3	check_sign()/give_way_sign()/ given_way	Action/Percept/Add Belief
watch_out_for_RU	Loc	watch()	Action
watching!	UA1	watching()	Action
busy?/RU_crossing?	UA1-UA2/UA2-UA4	road_user()/busy_roadjunction	Percept/Add Belief
wait?	UA2	wait/waiting(road_user)	Action/Percept
try_again!	UA1	try_again()	Percept
free?/RU_away?	UA1-UA2/UA2-UA4	no_road_user()/free_roadjunction	Percept/Add Belief
check_for_safe_gap	Loc	check_safe_gap()	Action
should_wait?/no_safe_gap?	UA1-UA2/UA2-UA5	no_safe_gap()/checking()	Percept/Action
enter?/yes_safe_gap?	UA1-UA2/UA2-UA5	safe_gap() or for_safe_gap()/enter	Percept/Action
AV_entered_RJ	Loc	enter_roadjunction	Percept
AV_away!	UA1	away_from_roadjunction	Add Belief

The mapping presented is direct where a given element from the Model has a matching element in the implementation. An additional correspondence, (that is not shown in the previous table) is the one from the UPPAAL Automata `sign`, `road_user`, and `safe_gap`, which are mapped to the random generation of these three events in the agent's environment.

However, not all elements can be mapped between the model and the implementation. For example, the `AV_agent` UPPAAL Automaton uses clock constraints that do not have a corresponding element in the agent's implementation. In addition, the GWENDOLEN code also has some details which are abstracted away in the timed model. The AV-agent has specific plans for different road junction rules (see the goals `enter_roadjunction_rules170_171` and `enter_roadjunction_rules170_172` in Listing 1). In this way, it is possible to keep track of which rules have been selected by the AV-agent.

Listing 1. AV-agent plans.

```
: Plans :

+! at_roadjunction (X,Y) [achieve] : { B av_away(0,0), B roadjunction(X,Y) }
    <- approach_roadjunction (X,Y);

+at_roadjunction (X,Y) : { B sign (Z,W) } <- check_sign (Z,W);

+stop_sign (Z,W) : { B sign (Z,W) }
    <- +stopped, +! enter_roadjunction_rules170_171 [perform];

+give_way_sign (Z,W) : { B sign (Z,W) }
    <- +given_way, +stopped, +! enter_roadjunction_rules170_172 [perform];

+! enter_roadjunction_rules170_171 [perform] :
    { B at_roadjunction (X,Y), B stopped, B to_watch (S,T) }
    <- watch (S,T);

+! enter_roadjunction_rules170_172 [perform] :
    { B at_roadjunction (X,Y), B given_way, B stopped, B to_watch (S,T) }
    <- watch (S,T);

+for_road_users (S,T) : { B road_user (S,T) }
    <- +busy_roadjunction, wait;
```

```
+waiting(road_user) : { B road_user(S,T) }
   <- watching(S,T);

+for_road_users(S,T) : { B no_road_user(S,T) }
   <- +free_roadjunction , check_safe_gap(S,T);

+try_again(S,T) : { B no_road_user(S,T) }
   <- +free_roadjunction , -busy_roadjunction , check_safe_gap(S,T);

+safe_gap(S,T) : { B no_road_user(S,T) }
   <- enter;

+no_safe_gap(S,T) : { B no_road_user(S,T) }
   <- checking(S,T);

+for_safe_gap(S,T) : { B new_safe_gap(S,T), B no_road_user(S,T) }
   <- enter;

+enter_roadjunction : { True }
   <- +away_from_roadjunction , done;
```

5.3. Implementation Details

Here, the implementation details concerning the AV-agent plans written in GWENDOLEN and the agent environment are described. Listing 1 presents a fragment of the agent implementation.

The first plan of the agent is designed to make the agent approach the RJ, so the action `approach_roadjunction(X,Y)` is invoked in the environment. This will only happen when the agent acquires the goal (of type achieve) `at_roadjunction(X,Y)` and has as guards the two beliefs: `av_away(0,0)` and `roadjunction(0,0)`. Next, the action `check_sign(Z,W)` is called, this action uses a random procedure to generate one of two possible outputs for traffic sign: stop or give way sign. To run this action the agent needs to perceive that it is `at_roadjunction(X,Y)` and believe there is a sign at (Z, W).

According to the existing traffic sign, a specific plan will be triggered. With this, we could track which RJ rule has been used by the agent. Either way, the agent will eventually call the action `watch(S,T)`, which is responsible for watching for road users. This action will return one of the two perceptions (from the environment): there is a road user or there is no road user.

In case there is a road user (i.e., exists a belief `road_user(S,T)`), the actions `wait` and `watching(S,T)` are executed. The former action will trigger a delay and the latter action is responsible for making the agent watch again for road users. The action `watching(S,T)` uses a random generation of road users at the road junction, in a way that at some point the road junction is supposed to be free of road users.

In case there is no road user (i.e., exists a belief `no_road_user(S,T)`), the agent believes the road junction is free and the action `check_safe_gap(S,T)` is executed. This action works similarly to action `check_sign(Z,W)` because it also uses a random generation to determine whether (or not) there is a safe gap at the road junction.

If there is no safe gap (i.e., exists a perception `no_safe_gap(S,T)`), a new action `checking(S,T)` is invoked, this action works similarly to action `watching(S,T)`, since it also uses random generator until at some point a safe gap arises at the road junction. Notice that to run action `checking(S,T)` the agent should still belief that there is no road user.

If there is a safe gap (and knowing that there is no road user), then the AV-agent may successfully enter the road junction. Once the agent has entered, it acquires a perception `enter_roadjunction`. After that, a new belief is added to the agent, so the agent knows that now it is away (once more) from the road junction (`away_from_roadjunction`).

5.4. Testing Scenarios

To test the SAE-RoR implementation stage we have run three different scenarios (see Figure 7). The setup of these scenarios corresponds to the placement of the road users in the road junction environment, which are the positions the AV-agent is supposed to watch for.

1. There are three road users, all at target spots, $\{(1,0); (1,1); (1,2)\}$.
2. There is one road user at a target spot, $(1,0)$. And two road users at safe spots, $\{(2,0); (2,2)\}$.
3. There are three road users, all at safe spots, $\{(0,0); (2,0); (2,2)\}$.

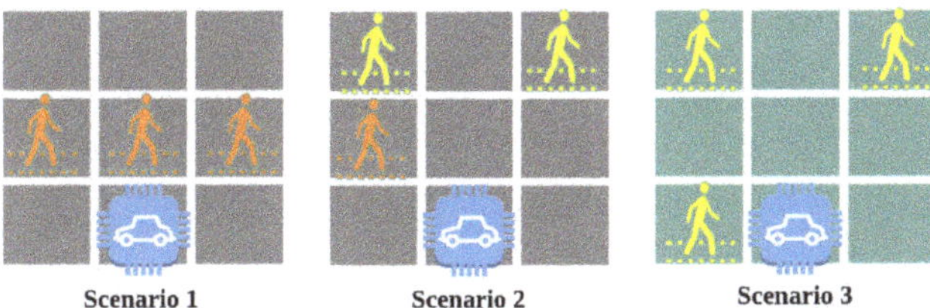

Figure 7. Three testing scenarios.

Figure 8 shows the output log from scenario 2. Notice that rule 171 is selected by the agent and action `watching(1,0)` is executed until the road junction is free, similarly, action `checking(1,0)` is also executed until there is a safe gap and the AV-agent may enter the road junction.

```
<terminated> run-AIL [Java Application] /usr/lib/jvm/java-8-openjdk-amd64/bin/java (Feb 23, 2021, 2:56:05 PM – 2:56:05 PM)
MCAPL Development Version 2019
ail.mas.DefaultEnvironment[INFO|main|2:56:05]: av done approach_roadjunction(0,1)
Stop sign (rule 171).
ail.mas.DefaultEnvironment[INFO|main|2:56:05]: av done check_sign(0,2)
AV has stopped
Busy at road user 1 position.
Free at road user 2 position.
Free at road user 3 position.
ail.mas.DefaultEnvironment[INFO|main|2:56:05]: av done watch(1,0)
ail.mas.DefaultEnvironment[INFO|main|2:56:05]: av done wait
keep watching until road junction is free
Road junction is free!
ail.mas.DefaultEnvironment[INFO|main|2:56:05]: av done watching(1,0)
AV has watched again
Is there a safe gap? false
ail.mas.DefaultEnvironment[INFO|main|2:56:05]: av done check_safe_gap(1,0)
No safe gap. Check again
Now, is there a (new) safe gap? true
ail.mas.DefaultEnvironment[INFO|main|2:56:05]: av done checking(1,0)
ail.mas.DefaultEnvironment[INFO|main|2:56:05]: av done enter
```

Figure 8. Scenario 2: output log.

As outlined above, the environment implementation has some actions that use random generation of events. As a result, we have run for each one of the three scenarios four different instances, this is necessary to properly capture all the possible outcomes of the actions. Specifically, we have observed the following elements:

- which RJ rule has been selected: rule 171 or 172.
- whether the RJ initially is busy or free.
- if initially the RJ has a safe gap or there is no safe gap.
- and whether or not the AV-agent has entered the RJ.

6. Formal Verification Results

In this section, we present the obtained results of our double-level model checking technique, where we have applied formal verification at design (using UPPAAL) and development (using MCAPL) levels.

All simulations and verifications (presented in this section) were done using the following specification: OS: *Linux Mint 19.3*; Processor: *Intel i7-8550U*; RAM: *8 GB*. And the correspondent software versions: UPPAAL *4.1.24—Academic*; MCAPL *development version*

2019 (NB: Our repository is available at https://github.com/laca-is/SAE-RoR, accessed on 23 June 2021).

We have successfully verified 30 properties (18 using UPPAAL and 12 using AJPF). These properties intend to capture all possible scenarios *w.r.t* the agent's behaviour against the three road junction rules (rules 170, 171, 172). With this, we intend to verify whether or not the agent is respecting the traffic rules according to the existing artefacts in the road junction environment.

6.1. Verification of Properties with UPPAAL

Below, we list the 18 properties written in TCTL that were successfully verified using UPPAAL. NB: `AV.x` represents the clock used in time constraints for the `AV_agent` automaton.

```
p1: A[] not deadlock
```

Description: a safety property which verifies if there is no deadlock.

```
p2: A[] ((RoadJunction.send_stop || RoadJunction.send_stop_and_give_way)
imply AV.AV_at_RJ)
```

Description: For all paths always the Road Junction environment when sending the AV to stop or to stop and give way to traffic, then the AV will be at the Road Junction.

```
p3: A[] (RoadJunction.waiting_for_AV imply A<> AV.watch_out_for_RU)
```

Description: For all paths always when the Road Junction is waiting for the AV, then for all paths at some time the AV watches out for road users.
NB: for the sake of clarity we use TCTL notation for this property, see in Section 3.1 the corresponding UPPAAL notation.

```
p4: A[] (AV.AV_check_for_safe_gap imply (RoadJunction.check_for_safe_gap ||
RoadJunction.there_is_safe_gap || RoadJunction.there_is_no_safe_gap))
```

Description: For all paths always when the AV checks for safe gap, then the Road Junction will be checking for a safe gap or it will know if (or not) there is a safe gap.

```
p5: A[] (RoadJunction.AV_may_enter imply A<> AV.AV_entered_RJ)
```

Description: For all paths always when the Road Junction tells the AV that it may enter the junction, then for all paths at some time the AV will enter the junction.
NB: the same remark for p3 is valid for p5.

```
p6: A[] (AV.AV_entered_RJ imply AV.x >= 2)
```

Description: For all paths always when the AV enters the Road Junction the clock (x) has a value greater or equal than 2.

```
p7: A[] ((AV.watch_out_for_RU) imply (AV.x >= 0 && AV.x <= 3))
```

Description: For all paths always the when the AV watches for Road Users at the Road Junction the clock (x) has a value between 0 and 3.

```
p8: A[] ((AV.waiting) imply (AV.x >= 1 && AV.x <= 5))
```

Description: For all paths always when the AV waits at the Road Junction the clock (x) has a value between 1 and 5.

```
p9: A[] ((AV.AV_check_for_safe_gap) imply (AV.x >= 2 && AV.x <= 4))
```

Description: For all paths always the when the AV checks for a safe gap at the Road Junction the clock (x) has a value between 2 and 4.

```
p10: A[] ((AV.watch_out_for_RU) imply
(RoadUser1.RU_crossing_RJ || RoadUser1.RU_away_from_RJ))
```

Description: For all paths always when the AV watches for (a single) road user, then it is only possible to have the road user crossing or away from the junction.

```
p11: A[] (RoadUser1.RU_crossing_RJ imply (RoadJunction.is_RJ_free || RoadJunction.busy_RJ
|| RoadJunction.AV_should_wait || RoadJunction.AV_is_waiting || RoadJunction.check_RU))
```

Description: For all paths always when there is a (single) road user crossing the junction, then it is only possible the Road Junction (environment) is checking for a road user or it is waiting or it should wait or it knows the junction is busy or yet it should check if the junction is free.

```
p12: A[] (RoadUser1.RU_crossing_RJ imply (not AV.AV_entered_RJ))
```

Description: For all paths always when there is a (single) road user crossing the junction, then it is not possible that the AV will enter the junction.

```
p13: A[] ((AV.watch_out_for_RU) imply ((RoadUser1.RU_crossing_RJ ||
RoadUser1.RU_away_from_RJ) || (RoadUser2.RU_crossing_RJ || RoadUser2.RU_away_from_RJ)))
```

Description: this is basically the same property as p10, except that here there are two Road Users at the Road Junction.

```
p14: A[] ((RoadUser1.RU_crossing_RJ || RoadUser2.RU_crossing_RJ) imply
(RoadJunction.is_RJ_free || RoadJunction.busy_RJ || RoadJunction.AV_should_wait ||
RoadJunction.AV_is_waiting || RoadJunction.check_RU))
```

Description: this is basically the same property as p11, except that here there are two Road Users at the Road Junction.

```
p15: A[] ((RoadUser1.RU_crossing_RJ || RoadUser2.RU_crossing_RJ) imply
(not AV.AV_entered_RJ))
```

Description: this is basically the same property as p12, except that here there are two Road Users at the Road Junction.

```
p16: A[] ((AV.watch_out_for_RU) imply ((RoadUser1.RU_crossing_RJ ||
RoadUser1.RU_away_from_RJ) || (RoadUser2.RU_crossing_RJ || RoadUser2.RU_away_from_RJ)
|| (RoadUser3.RU_crossing_RJ || RoadUser3.RU_away_from_RJ)))
```

Description: this is basically the same property as p10 and p13, except that here there are three Road Users at the Road Junction.

```
p17: A[] ((RoadUser1.RU_crossing_RJ || RoadUser2.RU_crossing_RJ ||
RoadUser3.RU_crossing_RJ) imply (RoadJunction.is_RJ_free || RoadJunction.busy_RJ ||
RoadJunction.AV_should_wait || RoadJunction.AV_is_waiting || RoadJunction.check_RU))
```

Description: this is basically the same property as p11 and p14, except that here there are three Road Users at the Road Junction.

```
p18: A[] ((RoadUser1.RU_crossing_RJ || RoadUser2.RU_crossing_RJ ||
RoadUser3.RU_crossing_RJ) imply (not AV.AV_entered_RJ))
```

Description: this is basically the same property as p12 and p15, except that here there are three Road Users at the Road Junction.

In Table 4 the execution results from the properties are summarised considering the existence (or not) of road users in the scenarios as well as the time and memory used to run each set of properties. Notice the highest values for time and memory respectively are 0.01 s and 49,396 KB, which can be seen as fair values even when three road users are considered in the simulation.

Table 4. Properties verified with UPPAAL—Execution results.

Properties	Scenario	Time	Memory
p1–p12	with 0 or 1 RU	0 s to 0.003 s	5800 KB/49,396 KB
p13–p15	with 2 RU	0.001 s	6040 KB/48,322 KB
p16–p18	with 3 RU	0.001 s to 0.01 s	6040 KB/48,322 KB

Table 5 presents the results from the 18 properties checked using UPPAAL. In this table we have classified each property according to the following:

- Road users: no road user at all; one, two, or three road users.
- System properties: two kinds of system properties are considered: temporal correctness and liveness.
- Interaction: that is those properties that present some sort of interaction with the environment.
- Quality: there are two kinds of properties related to quality: security and safety.
- Related Road Junction rules: each property is identified with the correspondent Road Junction rules that are related to the verified property.

Table 5. Properties verified with UPPAAL—Classification.

Property #	Road Users?	System Properties		Interaction	Quality		Related Rules		
		Temporal Correctness	Liveness	Interaction w/Environment	Safety	Security	R. 170	R. 171	R.172
p1	-					●			
p2	-			●	●			●	●
p3	-		●	●			●		
p4	-			●	●		●	●	
p5	-		●	●			●		
p6	-	●					●		
p7	-	●					●		
p8	-	●					●		
p9	-	●					●	●	
p10	1			●	●		●		
p11	1			●	●		●		
p12	1			●	●		●		
p13	2			●	●		●		
p14	2			●	●		●		
p15	2			●	●		●		
p16	3			●	●		●		
p17	3			●	●		●		
p18	3			●	●		●		

To discuss the verified properties and results we highlight some issues, as follows.

1. Properties p1 to p5 are related to security, safety, liveness, and interaction. In addition, these properties verify some of the main actions of our model, i.e., when AV-agent watches out for a road user, checks for a safe gap and for a traffic sign as well as will enter the RJ.

2. Properties p6 to p9 are responsible for verifying the time constraints included in the AV-agent automaton. With this, we can check temporal correctness for the main actions in our model, i.e., enter the RJ, watch out for road users, wait at RJ, and check for a safe gap.
3. Properties p10 to p12 are safety properties used to verify the effect of having a single road user at the RJ in some related actions. These properties formally verify what to expect when the AV-agent watches out road users and also when there is a road user crossing the junction what is allowed (and not) to happen considering the existent actions in the RJ environment.
4. Properties p13 to p15 run the same kind of verification from the previous item, except here the scenario considers the existence of two road users.
5. Properties p16 to p18 run the same kind of verification from item 3, except here the scenario considers the existence of three road users.
6. Related RJ rules: 16 properties are related to rule 170, which is indeed a general road traffic rule handling different possibilities of when and how a vehicle may enter the road junction. Moreover, rules 171 and 172 are also verified in specific properties.

The verification of properties with UPPAAL generates important information for stakeholders. Firstly, it is possible to check the main actions that can be taken by an AV-agent at Road Junction. Secondly, the time constraints included in our model which, as noted, are left implicit in non-digital highway codes, were shown to be reasonable and so can form recommendations for a Digital Highway Code. Thirdly, the model is efficient at analysing the scenario with three road users, where there is no increase in the use of time and memory. As a result, we believe it would be feasible to analyse more complex road junction models with more than three road users. Lastly, we have assessed the use of three Road Junction rules from the RoR, where the main actions and artefacts of each rule have been modelled and formally verified.

6.2. Verification of Properties with AJPF

We present the twelve properties (and their corresponding descriptions) that have been successfully verified with AJPF. NB: these properties are labelled with ap (representing *AJPF Property*) to distinguish them from the properties previously presented.

```
ap1: (B(av,sign(0,2)) & B(av,stopped)) -> [] G(av, enter_roadjunction_rules170_171)
```

Description: when AV believes there is a sign at $(0,2)$ and it has stopped, then it always obtains the goal of entering the road junction using rules 170–171.

```
ap2: (B(av,sign(0,2)) & B(av,given_way) & B(av,stopped)) ->
[] G(av, enter_roadjunction_rules170_172)
```

Description: when AV believes there is a sign at $(0,2)$, it has given way and stopped, then it always obtains the goal of entering the road junction using rules 170–172.

```
ap3: [] (B(av, at_roadjunction(1, 0)) -> <> (B(av, road_user(1, 0)) ||
B(av, no_road_user(1, 0)))
```

Description: It is always the case that if the AV is at a road junction at $(1,0)$, then eventually it will believe that either there is a road user at the junction at $(1,0)$ or there is not a road user at the junction at $(1,0)$.

```
ap4: [] (D(av,wait) -> (B(av,road_user(1,0)) & B(av,busy_roadjunction)))
```

Description: It is always the case that if the AV waits at the junction, then it believes there is a road user at $(1,0)$ and the road junction is busy.

```
ap5: [] ((B(av,no_road_user(1,0)) & B(av,free_roadjunction)) -> <> (B(av,no_safe_gap(1,0)
|| B(av,safe_gap(1,0)) || B(av,new_safe_gap(1,0)) || B(av,try_again(1,0))))
```

Description: It is always the case that when the AV believes there is no road user at $(1,0)$ and the road junction is free, then eventually the AV will acquire the belief there is no safe gap at $(1,0)$ or there is a safe gap (or a new safe gap) at $(1,0)$ or the belief it has tried again at $(1,0)$ (in the search for road users).

```
ap6: [] (D(av,check_safe_gap(1,0)) -> ~B(av,busy_roadjunction))
```

Description: It is always the case that if the AV·checks for safe gap at $(1,0)$, then it should not believe there is a busy road junction.

```
ap7: [] (D(av,check_safe_gap(1,0)) -> ~B(av,road_user(1,0)))
```

Description: It is always the case that if the AV checks for safe gap at $(1,0)$, then it should not believe there is a road user at $(1,0)$.

```
ap8: [] (D(av,check_safe_gap(1,0)) ->
(B(av,no_road_user(1,0)) & B(av,free_roadjunction)))
```

Description: It is always the case that if the AV checks for safe gap at $(1,0)$, then it believes there is no road user at $(1,0)$ and the road junction is free.

```
ap9: [] (D(av,enter) -> ~B(av,busy_roadjunction))
```

Description: It is always the case that if the AV enters the junction, then it should not believe there is busy road junction.

```
ap10: [] (D(av,enter) -> ~B(av,road_user(1,0)))
```

Description: It is always the case that if the AV enters the junction, then it should not believe there is a road user at $(1,0)$.

```
ap11: [] (D(av,enter) -> ~B(av,try_again(1,0)))
```

Description: It is always the case that if the AV enters the junction, then it should not believe to try again (and watch for a road user) at $(1,0)$.

```
ap12: [] (D(av, enter) -> ( B(av, safe_gap(1,0)) || B(av, new_safe_gap(1,0))
& B(av, no_road_user(1, 0)))
```

Description: It is always the case that if the AV enters the junction, then it believes there is a safe gap at $(1,0)$ (or a new safe gap) and no road user at $(1,0)$.

Table 6 shows the results obtained when running the AJPF model checker. These results consider Scenario 2 (previously seen in Figure 7), where there are three road users at the RJ, one of them is at a target spot and two are at safe spots.

All properties can be classified as safety properties. Properties ap1 and ap2 are specifically used to verify the application of the RJ rules, rules 170–171 and rules 170–172.

The remainder of the properties (from ap3 to ap12) are responsible for verifying that the AV-agent performs key actions involved in the rules at appropriate points: that is to watch for road users, wait, check for a safe gap and enter the road junction. In Figure 9 the execution log of ap12 is shown. In the last lines from the execution log (just above the results section) we notice the AV-agent knows the road junction is free (i.e., there is no road user) and that there is a new safe gap in the junction. This figure also shows no errors detected, in the results section of the execution log. This means that this property has been successfully model checked.

Table 6. Properties verified with AJPF—Execution results.

Property #	Results	Elapsed Time	States	Search	Instructions	Max Memory (MB)	Loaded Code
ap1	no errors detected	00:00:08	new = 703, visited = 201, backtracked = 904, end = 8	MaxDepth = 131, constraints = 0	36040819	603	Classes = 367, methods = 5647
ap2	no errors detected	00:00:08	new = 703, visited = 201, backtracked = 904, end = 8	MaxDepth = 131, constraints = 0	38107153	899	Classes = 368, methods = 5668
ap3	no errors detected	00:00:10	new = 703, visited = 201, backtracked = 904, end = 8	MaxDepth = 131, constraints = 0	39758810	731	Classes = 365, methods = 5630
ap4	no errors detected	00:00:07	new = 703, visited = 201, backtracked = 904, end = 8	MaxDepth = 131, constraints = 0	33160957	598	Classes = 368, methods = 5669
ap5	no errors detected	00:00:09	new = 703, visited = 201, backtracked = 904, end = 8	MaxDepth = 131, constraints = 0	45156165	896	Classes = 366, methods = 5651
ap6	no errors detected	00:00:07	new = 703, visited = 201, backtracked = 904, end = 8	MaxDepth = 131, constraints = 0	30367275	601	Classes = 367, methods = 5648
ap7	no errors detected	00:00:07	new = 703, visited = 201, backtracked = 904, end = 8	MaxDepth = 131, constraints = 0	31561846	601	Classes = 367, methods = 5648
ap8	no errors detected	00:00:07	new = 703, visited = 201, backtracked = 904, end = 8	MaxDepth = 131, constraints = 0	34589802	598	Classes = 367, methods = 5648
ap9	no errors detected	00:00:06	new = 703, visited = 201, backtracked = 904, end = 8	MaxDepth = 131, constraints = 0	30047443	602	Classes = 364, methods = 5629
ap10	no errors detected	00:00:06	new = 703, visited = 201, backtracked = 904, end = 8	MaxDepth = 131, constraints = 0	31242014	601	Classes = 367, methods = 5648
ap11	no errors detected	00:00:07	new = 703, visited = 201, backtracked = 904, end = 8	MaxDepth = 131, constraints = 0	31145552	600	Classes = 367, methods = 5648
ap12	no errors detected	00:00:08	new = 703, visited = 201, backtracked = 904, end = 8	MaxDepth = 131, constraints = 0	37515112	605	Classes = 368, methods = 5669

Figure 9. Execution log of ap12.

Considering the obtained results (seen in Table 6), we highlight the following: (i) all properties have beensuccessfully verified; ii properties took from 6 (ap9 and ap10) to 10 (ap3) seconds; (iii) the results related to the states, search space, and loaded code are basically the same for all properties; (iv) the number of instructions ranges from 30.367.275 (lowest value by ap6) to 45.156.165 (highest value by ap5); and (v) the amount of memory (in MB) ranges from 598 (ap4 and ap8) to 899 (ap2). The similarity of the results is a consequence of the fact that most of the computation effort in AJPF is related to the production of an automata that represents the implemented program [32] which is identical in all cases here.

The formal verification with AJPF acknowledges and offers some addition to the previous verifications (carried out with UPPAAL). Firstly, we successfully verify that the main actions the AV-agent can take at the RJ (watch, wait, check for a safe gap, and enter) are indeed invoked by the agent. Secondly, some properties include actions (e.g., `check_safe_gap`) that are only invoked in some cases (represented by the use of random in the environment) and so we also verify that these actions are taken when needed. Thirdly, we observed fair results in time, memory and other features obtained by the AJPF execution. At last, the verification process produces traces (and if necessary counter-examples) which allow us to identify which rules and random actions have been (autonomously) selected by AV-agent in any given scenario. This can be helpful, for example, if a scenario we verify leads to an accident, allowing a stakeholder to check and traceback the actions taken by the agent that led to that outcome and so advise on whether the agent, or possibly the representation of the rules in a Digital Highway Code, need to be amended.

7. Related Work

Here we analyse related work on the following topics: an AV application scenario, the Rules of the Road, some kind of Formal Verification technique (mainly Model Checking), some specification logic and the use of agents. Most of the works described here have as a goal the formal verification of a model related to AV.

In ref. [33], Luckcuck et al. present a survey on formal specification and verification of autonomous robotic systems. A number of these [34–37] apply formal verification to AVs, but none relate to our particular question around the design of Digital Highway Code rules that are intended to conform to pre-existing "Rules of the Road".

Table 7 summarises a comparison among the related work that is presented in the remainder of this section. The first three works [10,11,38] present some sort of formalisation for the road traffic rules (just like our approach does). Some interesting elements from these works are, correspondingly, the codification of traffic rules [10]; the solution for conflicts in traffic rules using a deontic logic [11]; the use of a real traffic data-set [38]. However, neither approach uses an agent abstraction to represent an AV decision-making. Kamali et al. and Al-Nuaimi et al. [12,13] include the formal verification (using Model Checking techniques) of BDI agents. But, their AV application scenario is not related to the road traffic rules.

Besides, Table 7 outlines some specific information concerning:

- Amount of road traffic rules used: some works (including our approach) represents 3 rules, but none represents more than this.
- Formal Verification tools: Ref. [12] and our work are the only ones that use two verification techniques at two different levels: design and development, in the other works a single technique is applied.
- Verification of properties: most properties are related to safety issues, but some include conflicts and consistency checking. Moreover, ref. [12] verifies 12 properties, ref. [13] 7, and ref. [10] 5 properties, while in our approach 30 properties are verified.
- Formalisation: all works use some kind of formalisation, most use temporal or deontic logic.
- Simulation tools: References [11–13] present the use of some graphical tool for simulation, which contribute for testing the system. Our approach uses the UPPAAL graphical tool for simulation, but for the agent's simulation we only use a cli (*command line interface*) tool.

NB: all works described in this table share the same goal of using a formalisation technique to represent an AV, where either the road traffic rules are formalised or some formal verification of agents is used.

Table 7. Related work comparison.

Work Reference	AV Application Scenario	Road Traffic Rules	Amount of Rules	Formal Verification Tools	Verified Properties (Type/Amount)	Formalisation (Logic)	Agent Programming	Simulation and Assessment
[10]	urban traffic/ lane changes	Yes. Overtaking rules (German)	3	Isabelle/HOL Theorem Prover	safe distance/5 (theorems)	LTL	No	Uses Isabelle's code generator to codify the rules in Standard ML.
[11]	urban traffic/ lane changes	Yes. Overtaking rules (Australia)	1 (rule 141)	Turnip (DDL reasoning tool)	exceptions and conflicts/not specified	DDL	No	Uses CARRS-Q driving simulator to generate experiment data; w/4 different scenarios; find legal and illegal driving behaviour; help of domain experts; conducted 24 experiments.
[38]	urban traffic/ lane changes	Yes. Safe distance between vehicles (Vienna convention)	1	No	safe distance/ not specified	algebraic equations	No	Uses real traffic data-set (NGSIM project, US Highway 101) on position, speed, acceleration, and lane of vehicles. Simulates safe and unsafe lane changes.
[12]	AV platooning	No	-	UPPAAL/ MCAPL/AJPF	safety and liveness/12	TCTL/LTL	Yes. BDI Agent.	Uses TORCS (car simulator) for environment simulation; a physical engine is implemented in MATLAB.
[13]	parking lot	No	-	MCMAS	stability and consistency/7	CTL	Yes. BDI Agent.	A graphical environment is created using ROS and Gazebo.
Our approach	urban traffic/ road junction	Yes. Road Junction rules (UK)	3 (rules 170/ 171/172)	UPPAAL/ MCAPL/AJPF	safety/30	TCTL/LTL	Yes. BDI Agent.	Uses UPPAAL and AJPF tools for simulation, w/3 different scenarios. Simulates the use of road junction rules by the AV.

7.1. Formal Verification of Agents

Kamali et al. [12] use same tools used in our work, UPPAAL, AJPF, and GWENDOLEN to model, implement and verify a vehicle platooning protocol. They use a mixed strategy that combines results from UPPAAL and AJPF, to deduce properties both of individual agents in the platoon and overall platoon behaviour. Our approach uses the two Model Checkers separately in order to verify properties of a single agent following proposed Digital Highway Code rules at different levels of abstraction.

In our architecture, the model developed in UPPAAL is used as a design template for the lower-level agent implementation. Thus, our model checking stages are loosely coupled, which is beneficial for modularity, allowing, for instance, the design level UPAAL model to be implemented in a different programming language.

Al-Nuaimi et al. [13] use Agent Model Checking to explore the behaviour of an AV in a parking lot. Their toolchain consists of the MCMAS model checker, Jason agent programming language, and CTL to verify temporal properties. The authors formally verify the AVs decisions. 12 rules are defined to verify planning, navigation, object detection and obstacle avoidance. ROS and Gazebo [39] are used to graphically simulate the application scenario. Again this work is targeted at the verification of proposed AV implementations from a safety perspective rather than in terms of the digitisation of rules of the road and verifying whether some agent can obey them.

7.2. The Formalisation of the Rules of the Road

Pek et al. [38] formalise the safety of lane change manoeuvres to avoid collisions. The authors use as reference the Vienna Convention on traffic rules to formalise a single rule on the safe distance.

Rizaldi et al. [10] formalise and codify part of the German Highway Code on the Overtaking traffic rules in LTL. They show how the LTL formalisation can be properly used to abstract concepts from the traffic rules and obtain unambiguous and precise specification for the rules. In addition, they formally verify the traffic rules using Isabelle/HOL theorem prover and also monitor an AV applying a given traffic rule, which has been previously formalised using LTL.

Bhuiyan et al. [11] assess driving behaviour against traffic rules, specifically the Overtaking rules from the Queensland Highway Code. Two types of rules are specified: overtaking to the left and the right. Moreover, they intend to deal with rules exceptions and conflicts in traffic rules (this is solved by setting priorities among the rules). Using DDL (Defeasible Deontic Logic) they assess the driving behaviour telling if the driver has permission or it is prohibited to apply a given rule for overtaking. The results basically show if the proposed methodology has recommended (or not) the proper behaviour for the driver (permission or prohibition). In addition, CARRS-Q, a driving simulator is used and 24 experiments are conducted in four different scenarios.

Our approache share the same goal: assessing AV behaviour against traffic rules (in our case, the road junction rules). However, we are using an agent-based implementation and verification, where it is also possible to tell when and how a given road junction has been selected and applied by the agent. In addition, our double-level model checking architecture results in the formal verification of 30 properties (18 at design and 12 at development level), which brings a comprehensible set of verification that ranges from time constraints properties (at design level) to specific actions that can (or can not) be taken by the AV in the road junction scenarios (at development level). To the best of our knowledge, [10,11,38], do not present this variety of abstraction levels in the properties they verify.

8. Conclusions

In Section 1 we have introduced three questions that we wanted to answer (i) Can these three selected road junction rules be used directly (i.e. as seen in the Highway code) by an AV? (ii) How to assess the AVs behaviour against the three road junction rules considering simple Road Junction scenarios? and (iii) Are there any guidelines that can be given to enable the AV to work correctly with such Road Junction rules?

The first question is answered by the formalisation and modelling proposed in our work. The Road Junction rules were abstracted and formalised into a Digital Highway code to render them machine-readable. To do this, it was necessary to remove ambiguity and make the rules explicit to the computational system. In addition, some degree of abstraction was necessary to handle similar terms as a single one, for example the term `safe-gap`, used in our work, can be found described in different ways throughout the rules from the UK Highway code.

The second question is answered by the own use of the double-level Model Checking technique and adoption of the methodology of exploring scenarios via random events from [18]. This generation of events makes it possible to simulate different scenarios within one model and explore all possible behaviours of the model's environment. By using the SAE-RoR architecture we have formally verified 30 properties (18 at the design level and 12 at the development level), these properties include security, safety, liveness, and temporal correctness properties, among others. We have obtained fair results considering the resources used (i.e., memory, time, search space, etc) in the verification of the properties (where all of these properties have been successfully verified). By running the verification of properties in the road junction simulated environment, we are able to capture and assess the AVs behaviour considering all possible actions (e.g., watch out, wait, check for safe gap, enter, etc) that can be taken by the AV-agent according to the three implemented road junction rules. Note that, while we do not claim that the properties verified completely represent all the possibilities, we believe that verification stages such as these will be necessary for reliable and compliant AVs.

For the third question, clearly we need a principled way to represent road junction rules in a machine-readable format. As part of this we need to identify and reify implicit the time constraints that appear in human-readable rules of the road. Similarly, the use of a BDI agent programming languages and Program Model Checking helps generate traces of AV-agent behaviour and so identify when and how a given Road Junction rule was applied. This kind of information is potentially of use to stakeholders.

We return to the trade-off mentioned in Section 1. Can a Digital Highway Code can be created with few minor changes or are several adaptions are necessary? We can only give an answer considering the subset from Road Junction rules that we have implemented here. These three rules express their ideas in sufficient detail for formal and executable representation in an AV. However, the rules still need some adaptation. In future work, we intend to revisit this question and develop a more general answer.

Having established the SAE-RoR architecture and workflow, we could now add the remaining Road Junction rules from the UK Highway Code. These remaining rules are similar to those already implemented, the differences lie primarily in the artefacts and the perceptions generated in the environment. For example, to add the Road Junction rules 175 and 176, which deal with Traffic Lights, we would need to represent the traffic light as an artefact and the green, amber, and red light as perceptions. But, the actions `stop` at the red light and `follow` at the green light, for instance, would not differ that much from actions already implemented for the AV-agent, like `wait` and `enter`. This work would be needed for full implementation of an AV but will yield little further insight at the methodological level.

Of more interest would be to consider a different section of traffic rules from the UK Highway Code, for example, the Roundabout rules in order to add generality to the framework. Similarly we could consider the inclusion of a Highway Code from a different country. Of particular interest would be to investigate how an AV-agent would work when travelling between countries when it would need to switch to a different set of Rules of the Road. The agent paradigm also allows us to explore behaviour in environments where agents have different profiles. Our AV-agent is supposed to behave according to the Road Junction rules. But, what will happen if it interacts with agents that violate traffic rules and can this be modelled and verified? This extension would potentially introduce the need for the implementation and verification of communication and cooperation algorithms. Following this idea where we would have a multi-agent system, we notice that some other aspects offer an interesting view on how to extend the SAE-RoR architecture to consider the implementation and verification of AVs protocols. For instance, the topics of distributed traffic control [40], vehicle-to-vehicle and vehicle-to-infrastructure communication [41], and also agent-based IoT (*Internet of Things*) applications [42], however at this moment these lie outside the issue considered here of adherence to "The Rules of the Road".

Moreover, we could improve our abstract model from the road junction rules by defining an extension of the Multi-lane Spatial Logic, as seen in [23]. With this logic, we could extend our representation in a way not to only capture the temporal aspects from the road junction rules, but also the spatial elements. Perhaps, a proper approach to represent a safe gap in an urban traffic environment, for instance.

Lastly, we aim to augment the SAE-RoR architecture with an Ethical Agent responsible for monitoring and verifying an agent's behaviour with respect to the Rules of the Road, as discussed in [43].

Author Contributions: Conceptualization, G.V.A., L.D. and M.F.; Formal analysis, G.V.A. and L.D.; Funding acquisition, M.F.; Investigation, G.V.A.; Supervision, L.D. and M.F.; Writing—original draft, G.V.A.; Writing—review & editing, G.V.A., L.D. and M.F. All authors have read and agreed to the published version of the manuscript.

Funding: This research was partially funded in the UK by EPSRC project EP/V026801 (the Trustworthy Autonomous Systems Node in Verifiability) and by the Royal Academy of Engineering, under the Chair in Emerging Technologies scheme. The funders had no role in the design of the study; in the collection, analyses, or interpretation of data; in the writing of the manuscript, or in the decision to publish the results.

Informed Consent Statement: Not applicable.

Conflicts of Interest: The authors declare no conflict of interest.

References

1. Avary, M.; Dawkins, T. *Safe Drive Initiative: Creating Safe Autonomous Vehicle Policy*; World Economic Forum: Geneva, Switzerland, 2020.
2. Prakken, H. On the problem of making autonomous vehicles conform to traffic law. *Artif. Intell. Law* **2017**, *25*, 341–363. [CrossRef]
3. Alves, G.V.; Dennis, L.; Fisher, M. Formalisation of the Rules of the Road for embedding into an Autonomous Vehicle Agent. In Proceedings of the International Workshop on Verification and Validation of Autonomous Systems, Oxford, UK, 18–19 July 2018; pp. 1–2.
4. Alves, G.V.; Dennis, L.; Fisher, M. Formalisation and Implementation of Road Junction Rules on an Autonomous Vehicle Modelled as an Agent. In *International Symposium on Formal Methods*; Lecture Notes in Computer Science; Sekerinski, E., Moreira, N., Oliveira, J.N., Ratiu, D., Guidotti, R., Farrell, M., Luckcuck, M., Marmsoler, D., Campos, J., Astarte, T., et al., Eds.; Springer International Publishing: Cham, Switzerland, 2020; pp. 217–232. [CrossRef]
5. Philipp, R.; Wittmann, D.; Knobel, C.; Weast, J.; Garbacik, N.; Schnetter, P. *Safety First for Automated Driving*; Daimler AG: Stuttgart, Germany, 2019.
6. Law Commission, U. *Automated Vehicles: Summary of the Analysis of Responses to Consultation Paper 2 on Passenger Services and Public Transport*; Law Commission: London, UK, 2020.
7. The British Standards Institution. *PAS 1882 Data Collection and Management for Automated Vehicle Trials*; The British Standards Institution: London, UK, 2020.
8. Waymo. *Safety Report*; Waymo LLC: Mountain View, CA, USA, 2020. Available online: https://waymo.com/safety (accessed on 23 June 2021).
9. Department for Transport. Using the Road (159 to 203)—The Highway Code— Guidance—GOV.UK. 2017. Available online: https://www.gov.uk/guidance/the-highway-code/using-the-road-159-to-203 (accessed on 23 June 2021).
10. Rizaldi, A.; Keinholz, J.; Huber, M.; Feldle, J.; Immler, F.; Althoff, M.; Hilgendorf, E.; Nipkow, T. Formalising and Monitoring Traffic Rules for Autonomous Vehicles in Isabelle/HOL. In Proceedings of the 13th International Conference on Integrated Formal Methods, Turin, Italy, 20–22 September 2017; pp. 50–66. [CrossRef]
11. Bhuiyan, H.; Governatori, G.; Rakotonirainy, A.; Bond, A.; Demmel, S.; Islam, M.B. *Traffic Rules Encoding Using Defeasible Deontic Logic*; IOS Press: Amsterdam, The Netherland, 2020. [CrossRef]

12. Kamali, M.; Dennis, L.A.; McAree, O.; Fisher, M.; Veres, S.M. Formal Verification of Autonomous Vehicle Platooning. *Sci. Comput. Program.* **2017**, *148*, 88–106. [CrossRef]

13. Al-Nuaimi, M.; Qu, H.; Veres, S.M. Computational Framework for Verifiable Decisions of Self-Driving Vehicles. In Proceedings of the 2018 IEEE Conference on Control Technology and Applications (CCTA), Copenhagen, Denmark, 21–24 August 2018; pp. 638–645. [CrossRef]

14. Bakar, N.A.; Selamat, A. Agent systems verification: Systematic literature review and mapping. *Appl. Intell.* **2018**, *48*, 1251–1274. [CrossRef]

15. Dennis, L.A. *Gwendolen Semantics: 2017*; Technical Report ULCS-17-001; Department of Computer Science, University of Liverpool: Liverpool, UK, 2017.

16. Bengtsson, J.; Larsen, K.; Larsson, F.; Pettersson, P.; Yi, W. UPPAAL—A tool suite for automatic verification of real-time systems. In *Hybrid Systems III*; Alur, R., Henzinger, T.A., Sontag, E.D., Eds.; Number 1066 in Lecture Notes in Computer Science; Springer: Berlin/Heidelberg, Germany, 1996; pp. 232–243. [CrossRef]

17. Dennis, L.A.; Fisher, M.; Webster, M.P.; Bordini, R.H. Model Checking Agent Programming Languages. *Autom. Softw. Eng.* **2012**, *19*, 5–63. [CrossRef]

18. Dennis, L.A.; Fisher, M.; Lincoln, N.K.; Lisitsa, A.; Veres, S.M. Practical Verification of Decision-Making in Agent-Based Autonomous Systems. *Autom. Softw. Eng.* **2016**, *23*, 305–359. [CrossRef]

19. Koeman, V.J.; Dennis, L.A.; Webster, M.; Fisher, M.; Hindriks, K.V. The "Why Did You Do That?" Button: Answering Why-Questions for End Users of Robotic Systems. *Lect. Notes Comput. Sci.* **2019**, *12058*, 152–172. [CrossRef]

20. Alves, G.V.; Dennis, L.; Fernandes, L.; Fisher, M. Reliable Decision-Making in Autonomous Vehicles. In *Validation and Verification of Automated Systems: Results of the ENABLE-S3 Project*; Leitner, A., Watzenig, D., Ibanez-Guzman, J., Eds.; Springer International Publishing: Cham, Switzerland, 2020; pp. 105–117. [CrossRef]

21. Fernandes, L.E.R.; Custodio, V.; Alves, G.V.; Fisher, M. A Rational Agent Controlling an Autonomous Vehicle: Implementation and Formal Verification. *Electron. Proc. Theor. Comput. Sci.* **2017**, *257*, 35–42. [CrossRef]

22. Dresner, K.; Stone, P. A Multiagent Approach to Autonomous Intersection Management. *J. Artif. Intell. Res.* **2008**, *31*, 591–656. [CrossRef]

23. Schwammberger, M. An abstract model for proving safety of autonomous urban traffic. *Theor. Comput. Sci.* **2018**, *744*, 143–169. [CrossRef]

24. Herrmann, A.; Brenner, W.; Stadler, R. *Autonomous Driving: How the Driverless Revolution Will Change the World*, 1st ed.; Emerald Publishing: Bingley, UK, 2018.

25. Nigeria, H.C. Nigeria Highway Code—III. ROAD JUNCTIONS. 2019. Available online: http://www.highwaycode.com.ng/iii-road-junctions.html (accessed on 23 June 2021).

26. Fisher, M. *An Introduction to Practical Formal Methods Using Temporal Logic*; Wiley: Hoboken, NJ, USA, 2011.

27. Baier, C.; Katoen, J.P. *Principles of Model Checking (Representation and Mind Series)*; The MIT Press: Cambridge, MA, USA, 2008.

28. Bratman, M.E. *Intentions, Plans, and Practical Reason*; Harvard University Press: Cambridge, MA, USA, 1987.

29. Hindriks, K.V. Programming Rational Agents in GOAL. In *Multi-Agent Programming: Languages, Tools and Applications*; El Fallah Seghrouchni, A., Dix, J., Dastani, M., Bordini, R.H., Eds.; Springer: Boston, MA, USA, 2009; pp. 119–157.

30. Visser, W.; Havelund, K.; Brat, G.; Park, S. Model Checking Programs. In Proceedings of the 15th IEEE International Conference Automated Software Engineering (ASE), Grenoble, France, 11–15 September 2000; pp. 3–12.

31. Bordini, R.H.; Hübner, J.F.; Wooldridge, M. *Programming Multi-Agent Systems in AgentSpeak Using Jason (Wiley Series in Agent Technology)*; John Wiley & Sons, Inc.: Hoboken, NJ, USA, 2007.

32. Dennis, L.A.; Fisher, M.; Webster, M. Two-Stage Agent Program Verification. *J. Log. Comput.* **2018**, *28*, 499–523. [CrossRef]

33. Luckcuck, M.; Farrell, M.; Dennis, L.A.; Dixon, C.; Fisher, M. Formal Specification and Verification of Autonomous Robotic Systems: A Survey. *ACM Comput. Surv.* **2019**, *52*, 100:1–100:41. [CrossRef]

34. Althoff, M.; Althoff, D.; Wollherr, D.; Buss, M. Safety verification of autonomous vehicles for coordinated evasive maneuvers. In Proceedings of the 2010 IEEE Intelligent Vehicles Symposium, San Diego, CA, USA, 21–24 June 2010; pp. 1078–1083. [CrossRef]

35. Pallottino, L.; Scordio, V.G.; Bicchi, A.; Frazzoli, E. Decentralized Cooperative Policy for Conflict Resolution in Multivehicle Systems. *IEEE Trans. Robot.* **2007**, *23*, 1170–1183. [CrossRef]

36. Heß, D.; Althoff, M.; Sattel, T. Formal verification of maneuver automata for parameterized motion primitives. In Proceedings of the 2014 IEEE/RSJ International Conference on Intelligent Robots and Systems, Chicago, IL, USA, 14–18 September 2014; pp. 1474–1481. [CrossRef]

37. Kress-Gazit, H.; Wongpiromsarn, T.; Topcu, U. Correct, Reactive, High-Level Robot Control. *IEEE Robot. Autom. Mag.* **2011**, *18*, 65–74. [CrossRef]

38. Pek, C.; Zahn, P.; Althoff, M. Verifying the safety of lane change maneuvers of self-driving vehicles based on formalized traffic rules. In Proceedings of the 2017 IEEE Intelligent Vehicles Symposium (IV), Redondo Beach, CA, USA, 11–14 June 2017; pp. 1477–1483. [CrossRef]

39. Quigley, M.; Conley, K.; Gerkey, B.P.; Faust, J.; Foote, T.; Leibs, J.; Wheeler, R.; Ng, A.Y. ROS: An Open-source Robot Operating System. In Proceedings of the ICRA Workshop on Open Source Software, Kobe, Japan, 12–17 May 2009.

40. Bui, K.H.N.; Jung, J.J. Internet of agents framework for connected vehicles: A case study on distributed traffic control system. *J. Parallel Distrib. Comput.* **2018**, *116*, 89–95. [CrossRef]

41. Alouache, L.; Nguyen, N.; Aliouat, M.; Chelouah, R. Toward a hybrid SDN architecture for V2V communication in IoV environment. In Proceedings of the 2018 Fifth International Conference on Software Defined Systems (SDS), Barcelona, Spain, 23–26 April 2018; pp. 93–99. [CrossRef]
42. Savaglio, C.; Ganzha, M.; Paprzycki, M.; Bădică, C.; Ivanović, M.; Fortino, G. Agent-based Internet of Things: State-of-the-art and research challenges. *Future Gener. Comput. Syst.* **2020**, *102*, 1038–1053. [CrossRef]
43. Alves, G.V.; Dennis, L.; Fisher, M. First Steps towards an Ethical Agent for Checking Decision and Behaviour for an Autonomous Vehicle on the Rules of the Road. In Proceedings of the Second Workshop on Implementing Machine Ethics, Dublin, Ireland, 30 June 2020; Zenodo: Geneva, Switzerland, 2020; pp. 1–2. [CrossRef]

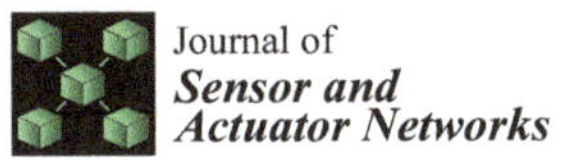

Journal of
**Sensor and
Actuator Networks**

Article

A Programming Approach to Collective Autonomy

Roberto Casadei *, Gianluca Aguzzi and Mirko Viroli

Department of Computer Science and Engineering (DISI), Alma Mater Studiorum–Università di Bologna,
47521 Cesena, Italy; gianluca.aguzzi@unibo.it (G.A.); mirko.viroli@unibo.it (M.V.)
* Correspondence: roby.casadei@unibo.it

Abstract: Research and technology developments on autonomous agents and autonomic computing promote a vision of artificial systems that are able to resiliently manage themselves and autonomously deal with issues at runtime in dynamic environments. Indeed, autonomy can be leveraged to unburden humans from mundane tasks (cf. driving and autonomous vehicles), from the risk of operating in unknown or perilous environments (cf. rescue scenarios), or to support timely decision-making in complex settings (cf. data-centre operations). Beyond the results that individual autonomous agents can carry out, a further opportunity lies in the collaboration of multiple agents or robots. Emerging macro-paradigms provide an approach to programming whole collectives towards global goals. Aggregate computing is one such paradigm, formally grounded in a calculus of computational fields enabling functional composition of collective behaviours that could be proved, under certain technical conditions, to be self-stabilising. In this work, we address the concept of collective autonomy, i.e., the form of autonomy that applies at the level of a group of individuals. As a contribution, we define an agent control architecture for aggregate multi-agent systems, discuss how the aggregate computing framework relates to both individual and collective autonomy, and show how it can be used to program collective autonomous behaviour. We exemplify the concepts through a simulated case study, and outline a research roadmap towards reliable aggregate autonomy.

Keywords: collective autonomy; self-organisation; aggregate computing; multi-agent systems; coordination

Citation: Casadei, R.; Aguzzi, G.;
Viroli, M. A Programming Approach
to Collective Autonomy. *J. Sens.
Actuator Netw.* **2021**, *10*, 27. https://
doi.org/10.3390/jsan10020027

Academic Editor: Rafael C. Cardoso,
Angelo Ferrando, Daniela Briola,
Claudio Menghi and
Tobias Ahlbrecht

Received: 15 March 2021
Accepted: 13 April 2021
Published: 19 April 2021

1. Introduction

Research and technology trends promote a vision of artificial systems that are able to resiliently manage themselves and autonomously deal with issues at runtime in dynamic environments. Such a vision is mainly investigated by two related research threads. One is the field of autonomic computing [1] and self-adaptive systems [2], which promote the development of Information and Communications Technology (ICT) systems able to self-manage given a set of high-level goals. Indeed, endowing systems with higher degrees of autonomy can be leveraged to unburden humans from mundane tasks (cf. driving and autonomous vehicles), from the risk of operating in unknown or perilous environments (cf. rescue scenarios), or to support timely decision-making in complex settings (cf. data-centre operations). The other is the field of multi-agent systems (MAS) [3], which evolved from the field of distributed artificial intelligence [4]. An MAS is a system of agents, i.e., a collection of autonomous entities interacting with their environment [5] and other agents to satisfy their design objectives. From an engineering point of view, agents and related abstractions are considered useful tools for the analysis and design of complex software-based systems. There are two key problems in the use and development of agents: the design of individual agents (micro level) and the design of a society of agents (macro level) [3].

Indeed, beyond the results that individual autonomous agents can carry out, a further opportunity lies in the collaboration of multiple agents or robots. Emerging macro-paradigms [6,7] provide an approach to programming whole collectives towards global goals. Aggregate computing or programming [7,8] is one such paradigm, formally grounded in a calculus of computational fields [9] (maps from agents to values) enabling functional specification and composition of collective behaviours. The major benefit of the aggregate programming approach is that it explicitly addresses collective adaptive behavior, rather than the behavior of individuals (which is addressed indirectly, as a consequence of the intended global behavior). The idea is to code scripts, conceptually executed by the collective as a whole, in terms of reusable building blocks of collective tasks [10] capturing both state, behavior, and interaction, crucially enjoying formal mapping to the behavior of individuals and often provable convergence properties [7,11]. We argue that this programming approach, which originated from the research areas of coordination [12] and spatial computing [6] (see [7] for a historical note), can be suitable to MAS programming. Accordingly, in this work, we address the concept of collective autonomy, i.e., the form of autonomy that applies at the level of a group of individuals. Though this notion has been investigated in the literature (cf. Section 2), there are mainly preliminary approaches for practical programming of collective autonomous behavior by a global perspective; so, in this manuscript, we address this software engineering problem explicitly, and sketch a roadmap for further research. Therefore:

- we provide a review of literature about autonomy and especially collective autonomy in MASs (Section 2);
- we analyze the aggregate computing framework by the perspective of autonomy, by covering its positioning with respect to individual and collective autonomy, and showing how it can support adjustable autonomy (Section 3);
- we exemplify the discussion through a simulated case study, investigating (i) the relationship between individual goals/autonomy and collective goals/autonomy; and (ii) the relationship between structures and collective autonomy (Section 4); and
- we discuss gaps in the literature on programming reliable collective autonomy and delineate a research roadmap (Section 5).

Finally, we conclude with a wrap-up in Section 6. In summary, in this route, we cover how to aggregate computing relates to and supports various forms of autonomy, and propose it as a framework for programming and simulating collective autonomous behavior, in a way that differs from related works and that opens up various research directions regarding actionable notions of collective autonomy.

2. Background and Related Work

This section provides the background for our contribution, which is positioned at the intersection of distributed artificial intelligence and software engineering.

2.1. Autonomy in Software Engineering and Multi-Agent Systems

At a first level, autonomy is used as a general informal notion, a characteristic assumed to be possessed by some entities during design. Etymologically, autonomy refers to an entity that follows its own laws, i.e., that is able to self-regulate its behavior. At a second level, concrete and (semi-)formal notions of autonomy are developed to support engineering tasks under specific viewpoints—see Table 1 for a summary. For instance, from a programming language perspective, agents are computationally autonomous in the sense that they "encapsulate invocation" (i.e., they act as internally defined, e.g., by rules or goals). Therefore, agents can be thought of as the next step of an evolution from monoliths to modules (encapsulation of behavior), to objects (encapsulation of state in addition to behavior), to active objects/actors (decoupling invocation from execution) [13].

Autonomy, together with agency (the ability to act), appears to one of the key defining and agreed upon characteristics of agents in MAS research [14]. From autonomy and agency, other features naturally arise. Agents are proactive: they are not only reactive to external stimuli, but driven towards action by an inner force. Agents are social and interactive, as autonomy makes sense in a relational context such as a society (MAS). The importance of interaction has motivated research on how to effectively rule it to promote the satisfaction of design goals—the field of coordination [12].

As a relational notion, it is more precise to say that an agent is autonomous *(i) from something* and possibly *(ii) with respect to something* [15]. Accordingly, it is possible to distinguish between social autonomy (the autonomy of an agent from other agents) and non-social, environmental autonomy (the autonomy of an agent from the environment). It is important to remember that autonomy is a gradable notion from the extremes of no autonomy to full or absolute autonomy. Regarding the object for which autonomy is considered, researchers typically distinguish between agents that are plan-autonomous (i.e., are free to determine the course of actions to reach given goals) and agents that are goal-autonomous (i.e., are free to determine their own goals). These forms are also called as executive and motivational autonomy [16], respectively. Another common distinction is between weak and strong agency. In the latter, goals are explicitly represented. An approach to (strong) agency is to consider agents as intentional entities with mental states such as epistemic (e.g., percepts, beliefs) and motivational (e.g., desires, intentions) states. A well-known model in this class is the Belief–Desire–Intention (BDI) control architecture, which counts several implementations and variants [17].

Table 1. A summary of common notions of autonomy.

Dimension	Elements/Terms	
Reference entity	Agent	Group of agents
	individual autonomy	collective autonomy
Autonomy "from" something	Other agents	Environment
	social autonomy	non-social/environmental autonomy
Autonomy "with respect to" something	Goals	Plans
	motivational autonomy	executive autonomy
Autonomy extremes	None	Full
	no autonomy (passivity)	absolute autonomy (freedom)
Autonomy flexibility	None	Full
	fixed autonomy	adjustable autonomy
Source of autonomy with respect to a component of a reference entity	Internal	External
	endogenous autonomy	exogenous autonomy

2.2. Collective Autonomy

The notion of collective autonomy emerges when the reference agent is not an atomically individual agent but a whole collective, i.e., a collection of individuals (agents or other, possibly non-autonomous agents). As autonomy as a concept tends to be related to the existence (and possibly awareness) of a "self" [18], an autonomous collective tends to be and work as a "unit". Working as a unit requires the components of a collective to be jointly directed towards goals, states of affairs, or values—a concept known as collective intentionality [19,20]. In [19], an (intentional) collective is defined as a collection of agents held together by a "plan", which specifies a "goal" and their "roles". In [21], a formal analysis of collective autonomy is provided. A collective (agent) is defined as a collection of complementary agents sharing a common, collective goal (which, in a sense, reduces the individual autonomy of the members). As for individuals, autonomy for collectives is a gradable and relational notion. A collective may be defined as plan-autonomous (goal-autonomous) if no other entity (internal or external) can change its plans (goals).

The focus on autonomy, together with a collective stance, can be used to model or engineer complex system behavior. This is the idea of autonomy-oriented computing [22], where computation is defined in terms of local autonomous entities, spontaneously (inter-)acting together and with the environment to achieve self-organizing behavior. In the following section, we cover a state-of-the-art programming model for this paradigm.

2.3. Multi-Agent Systems Programming

A recent survey on agent-based programming is by Mao et al. [23]. They classify agent programming languages into three families according to the level they address: individual agent programming (micro-level), agents integration and interaction programming (meso-level), and multi-agent organisation programming (macro-level). Representatives of these classes include cognitive-oriented languages such as AgentSpeak(L)/Jason [24], agent communication and environment modelling languages like KQML [25], CArtAgO [26], and SARL [27], and organisation-oriented programming such as MOISE [28]. Another recent survey by Cardoso et al. [29] distinguishes between general-purpose agent programming languages and languages for agent-based modeling and simulation. In the following, we cover a programming model arising from research on spatial computing [6,30] and field-based coordination [7,31], which provides an original approach to MAS programming that allows driving micro-level activity based on specifications addressing the meso- and macro-levels of a MAS.

2.3.1. Aggregate Programming

Aggregate programming is an approach to specify the collective adaptive or self-organising behaviour of a MAS by a global perspective. The individual behavior of the agents derives from an aggregate program that is conceptually executed by the system as a whole. The aggregate program provides a way to map the local observations of an individual agent (i.e., sensing information, current agent state, and inbound messages from neighbors) to (eventually) globally-coherent local actions (i.e., actuation instructions, and outbound messages). Therefore, an aggregate program covers the aspects of sensing, actuation, computation, and communication to define how the MAS should collectively behave. In particular, we define an aggregate system as a MAS of agents, structurally connected such that an agent can only interact with a subset of other agents known as its neighbors, and repeatedly plays an aggregate program against its up-to-date context (further details about the execution protocol are provided in Section 3.1).

Historically, aggregate programming originated from works drawing inspiration from nature: whereas the biological inspiration led to swarm intelligent MASs, where agent indirectly interact by pheromones [32], the physical inspiration led to the idea of agents acting in environments empowered with potential fields [31]. Recently, aggregate programming has been formally backed by field calculi [7], which provide a compositional approach to global behavior specification based on functions from fields to fields. A (computational) field is a map associating a value to any device of a given domain. So, for instance, controlling the movement of a swarm of drones can be expressed through a field of velocity vectors, which maps any drone of the swarm to a corresponding velocity (speed and direction); the set of low-energy devices can be denoted through a Boolean field holding `true` for devices whose local energy level (as perceived by local sensors, and collectively also denoted as a floating-point field) is under a certain threshold (also a floating-point field). These fields, then, are generally manipulated through three kinds of constructs:

1. Stateful evolution: `rep(init)(f)`—expressing how a field, starting as `init`, should evolve round-by-round through unary function `f`.
2. Neighbour interaction: `nbr(e)`—used to exchange with neighbours the value obtained by evaluating field expression `e`; this locally yields a neighbouring field, i.e., a field that maps any neighbour to the corresponding evaluation of `e`.

3. Domain partitioning: `branch(c){ifTrue}{ifFalse}`—used to partition the domain of devices into two parts: the devices for which field c is locally `true`, which evaluate expression `ifTrue`, and those for which c yields `false`, which evaluate `ifFalse`.

The idea of aggregate programming is to write programs talking about global behavior (fields) and let these drive the local activity of every device in the system. Aggregate programming is embodied by concrete aggregate programming languages [7], such as ScaFi [33,34], a Domain-Specific Language (DSL) embedded in Scala as well as a toolchain for aggregate system development and simulation [35]. ScaFi is used for the examples in this paper and for the experimental evaluation of Section 4.

We adopt ScaFi in this paper mostly for practical reasons: with respect to other aggregate programming languages such as Proto and Protelis, surveyed in [7], ScaFi is a strongly typed, internal DSL; therefore, it enables straightforward reuse of powerful features from the Scala host language (including its type system, type inference, programming abstractions, libraries) as well as seamless integration with tooling supports for Java Virtual Machine-based languages (including Integrated Development Environments, static analysis, and debugging tools), at the expense of a more constrained syntax and semantics. Additionally, ScaFi also represents an agile framework for testing experimental language features (cf. aggregate processes [34]—referenced in Section 5.1). Hence, among the existing languages for aggregate programming, we believe ScaFi is the one better fitting rich scenarios like those addressed in this paper.

A full account of research about aggregate programming, field calculi, and ScaFi is beyond the scope of this article; the interested reader can refer to [7,34].

Recently, some preliminary work [36] has been carried out to consider the application of the aggregate approach for MAS programming, along with a strong-agency viewpoint. There, two main ideas are proposed. One is the notion of a cognitive field (e.g., fields of beliefs, fields of goals, fields of intentions), which could be used to represent "a kind of distributed, decentralized, and externalized mental state". The other is the notion of an aggregate plan, i.e., a global, collective plan of actions modeling the way in which a dynamic team of agents cooperates towards a social goal in a self-organizing way. In particular, aggregate plans can be created by an initiator agent and iteratively spread by the other agents; any agent has the faculty of choosing to adopt the plan or not; if the plan is adopted, then the agent will execute the corresponding actions, which in general will depend on the agent's position in space and in the team. This mechanism may be suitable where the behavior to be executed is somehow related to the space/environment, where the MAS can be clustered in teams of agents exhibiting uniform behavior, or where the MAS has to embed a decentralized, self-organizing force. A management lifecycle for aggregate plans involve the following phases: synthesis, spreading, collection, selection, execution. In this article, we build on this perspective, and rather focus on the notion of (collective) autonomy.

3. Autonomy in Aggregate Computing

In this section, we analyse the notion of autonomy (cf. Section 2.1) by the aggregate computing and programming perspective (cf. Section 2.3.1). In particular, we propose the aggregate execution protocol as basis of an agent control architecture (Section 3.1), and then discuss aggregate programming of individual (Section 3.2) as well as collective autonomy (Section 3.3).

3.1. Aggregate-Oriented Agent Control Architecture

The field calculus small-step operational semantics [9] provides an abstract aggregate execution protocol for "driving" aggregate behaviour. The aggregate execution protocol typically consists of having an agent repeatedly run (e.g., once per second, or upon change of the local context) a computation round consisting of the following steps:

1. Context evaluation. In this step, the agent looks at its current state, its sensors, and its message box for new information in order to update the local context.

2. Aggregate program evaluation. In this step, the agent runs the aggregate program providing its local context as an input: the evaluation of the program returns an output data structure that can be used for context update.

3. Context action. Using the output of the program evaluation, the agent must

    *   update its local state;
    *   send a message to neighbours;
    *   trigger actuations (if needed).

Such an abstract aggregate execution protocol can effectively be used to define an agent control architecture (Figure 1). Variants of such a control architecture can also be envisaged, e.g., by considering asynchronicity and different rates for context evaluation, computation, and action. Preliminary work supporting this direction can be found in [37,38], where partitioning schemas and programmable schedulers are proposed.

This control architecture also provides a basis for integrating the aggregate paradigm with other agent control architectures, e.g., the cognitive ones based on BDI [17]—which makes for an interesting future fork.

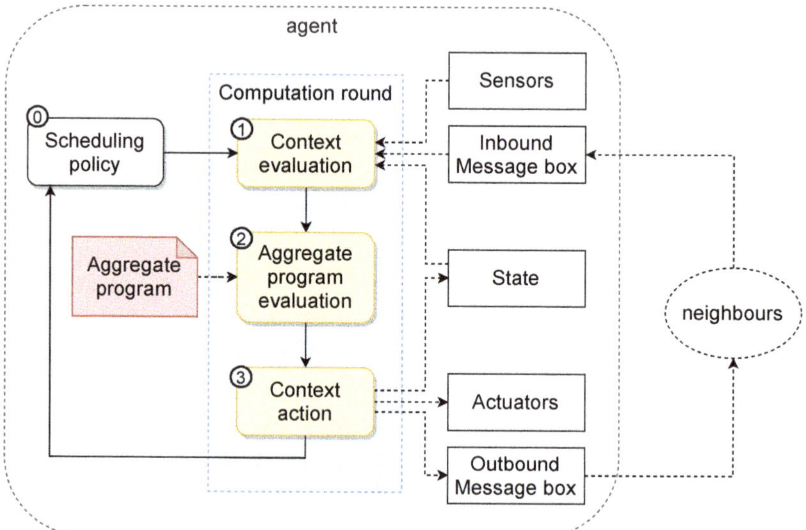

Figure 1. Agent control architecture in aggregate programming. Notation: square boxes denote components; rounded boxes denote activities; dashed rounded boxes denote agents; solid arrows denote control passing; dashed arrows denote data passing.

3.2. Individual Autonomy in Aggregate Computing

As a running example, consider a crowd detection and steering application [8] (cf. Figure 2a). When this application is built with the aggregate paradigm, the crowd is represented as an aggregate system consisting of a large-scale, dense network of smartphone- or wearable-augmented people—co-located in a spatial region such as a public exhibition area, a concert, or a stadium. An aggregate program can be continuously played by this system, providing each agent with an estimation of the local density, a local risk level, and possibly—if the vicinity to risky areas exceeds a certain threshold—advice for safe dispersal (in terms of a field of movement directions). In ScaFi, such a program may be implemented, reusing a library of general-purpose aggregate components [10], as per Figure 2c. Function `crowdTracking` (Line 2) runs a collective crowding risk estimation process, by calling `collectiveDensityEstimation` (Line 6), and selects the corresponding output (i.e., a crowding value—`NoRisk`, `Risk`, or `Danger`) only for the devices for which the perceived local density exceeds threshold value `thCrowd`; for the others, the output is `NoRisk`. Function `collectiveDensityEstimation` (whose code is not shown) may, e.g., using the building blocks in [10], break the system into multiple areas, compute the mean local density in each one of them, and share to all the members the area-wide crowding level (based on whether the average density exceeds threshold `thDanger`). Then, function `crowdDispersal` (Line 10) provides a suggestion for dispersal for all the devices that are closer than `riskRange` from any device with crowding level `Risk`; the suggestion is essentially computed as the opposite of a vector pointing to the center of mass of a close `Overcrowded` device group. We remark that, though the aggregate program is unique (essentially like a shared plan for all the involved agents), its execution is distributed (i.e., decentralized) and local (i.e., with agents interacting with neighbor agents only), hence enabling scalable collective computations.

An aggregate behavior is the result that emerges from the combination of (i) an aggregate program; (ii) a concrete aggregate execution; (iii) an environment—which comprises the behavior of the agents that is out of the control scope of the aggregate program. For the crowd example, the aggregate program expresses how the MAS should determine risky areas and how dispersal processes should be carried out; an aggregate execution may, e.g., set the round frequency to match the levels of mobility; the environment entails elements like spatial distribution and connectivity. Therefore, for an aggregate behavior to actually work, i.e., to correctly carry out its intended functionality (which can be seen as a "social benefit") it is important that the aggregate execution which is tailored to match prefigured features of the environment, is respected by every device, generally. Indeed, there exist guarantees regarding (i) self-stabilisation of aggregate computations [11], namely the guarantee to eventually converge to a correct state once perturbations cease; and (ii) adaptation of aggregate behaviours to device distribution (density, topology) [39], namely eventual consistency of values, whose approximation improves as the discrete network tends to more densely cover the environment. Under this perspective, an aggregate program may be interpreted as a representation of a social norm, and the aggregate system as a society, or even a normative MAS [40]. So, an agent that does not run the aggregate program is a deviant and not really part of society. On the other hand, playing by the norms implies a limitation of individual autonomy, which is generally traded for a greater, social good. In the crowd example, it is clear that the activity of information gathering and broadcasting is instrumental to achieve good "social" performance in risk detection and dispersal self-correction; moreover, great damage may result from not following dispersal advice—so, the crowd program may be interpreted as the reification of a social norm for safe behavior in a perilous situation. Inspired by normative MAS, deviance can be minimized by leveraging social enforcement mechanisms such as rewards or sanctions [40]. However, an agent of an aggregate MAS may reduce, e.g., the frequency at which rounds are executed, because of a low-energy level, or because its portion of the environment is largely stationary. In other words, the first element of individual autonomy in aggregate MASs revolves around how an agent adheres to the aggregate execution protocol for the application at hand.

The other elements of individual autonomy are those provided by and undergone by the aggregate program. Indeed, an aggregate program may be used to control individual behavior, through local functions containing no aggregate constructs for global coordination. We call this endogenous or delegated autonomy, as it is delegated from the (inside of the) program itself. For instance, the crowd example may be extended to leave some individual autonomy when following the dispersal advice:

```
localDispersalDecision(crowdDispersal(...))
```

where `localDispersalDecision` can be a function potentially different from agent to agent, able to affect the local socially-enforced dispersal direction. This mechanism may also be used to program adjustable autonomy, i.e., the form supporting "agents with graded autonomy properties" [41].

(**a**)   (**b**)

```
1   /* Shared plans (see also [8]) */
2   def crowdTracking(p: Double, r: Double, w: Double, t: Double,
3                   thCrowd: Double, thDanger: Double): Crowding = {
4     val localDensity = localDensityEstimation(p, r)
5     mux(recentlyTrue(localDensity > thCrowd, t)){
6       collectiveDensityEstimation(p, r, thDanger)
7     } { NoRisk }
8   }
9
10  def crowdDispersal(c: Crowding, r: Double): TargetPosition =
11      // if a device is closer than r to any device with crowding level = Risk or higher
12      // then compute a movement vector to move away from the Overcrowded area, or do not move
13      branch(distanceTo(c >= Risk) < r){ vectorFrom(c == Overcrowded) }{ currentPosition() }
14
15  /* Shared beliefs: see [8] for motivation of the specific values */
16  val p = 0.005 // proportion of people with a corresponding device (agent)
17  val r  = 30   // range in metres for local crowding estimation
18  val w  = 0.25 // fraction of walkable space
19  val t  = 60.0  // timeframe (in seconds) for risk monitoring
20  val thCrowd  = 1.08 // relevant crowding threshold
21  val thDanger = 2.17 // dangerous crowding threshold
22  val riskRange = 50   // distance to risk for triggering alert
23
24  /* Aggregate program: main logic */
25  val crowdingLevel = crowdTracking(p, r, w, thCrowd, thDanger, t)
26  crowdDispersal(crowdingLevel, riskRange) // dispersal advice
```

(**c**)

Figure 2. Crowd detection and steering example: snapshot, architecture and program. (**a**) A simulation snapshot of the crowd detection and steering example. Red nodes are devices in an overcrowded area; cyan nodes are nodes at risk since close to overcrowded devices; black nodes are in a safe location. Solid lines denote connectivity links (the longer links are those between infrastructural nodes); (**b**) diagram corresponding to the ScaFi application code below. Boxes denote computational field expressions; solid arrows denote input/output relationships; dashed arrows are a shorthand to denote conditional selection (when initially separated—for mux, where the "then" and "else" expressions are both evaluated, but only one is returned) and conditional evaluation (when initially joined—for branch, where only one between the "then" and "else" expression is evaluated and returned). (**c**) Aggregate program as implemented in ScaFi.

On the other hand, an aggregate program may abstract over or depend on certain elements of agent autonomy. For instance, in the crowd monitoring and control example, the people—and hence the corresponding digital twins (agents)—maintain autonomy regarding mobility (which may also be affected by the crowd itself), since they may choose to follow the dispersal advice or not. We call this exogenous autonomy, as it comes from the outside, beyond the control of the aggregate program. This also includes the influence that an agent may exert on the aggregate program by manipulating its inputs (i.e., state, sensors, and messages). Aggregate programs are typically developed to expressively deal with exogenous autonomous behavior, with strategies that the whole aggregate MAS can follow to adapt and overcome corresponding perturbations to the collectively desired state-of-affairs. Table 2 synthesises the various sources and forms of individual autonomy in aggregate computing.

Table 2. Individual autonomy in aggregate computing.

Individual Autonomy Element	**Range (Less–More)**
Adherence to the aggregate execution protocol	Fully autonomous (uncooperative)—Completely adherent (control-driven)
Endogenous autonomy (provided by the AC program)	Uncontrolled–Controlled
Exogenous autonomy (undergone by the AC program)	Controlled–Uncontrolled

3.3. Collective Autonomy in Aggregate Computing

As discussed in Section 2.2, collective autonomy is the form of autonomy exhibited by a collective, i.e., a MAS as a whole. Such a concept can be framed around two notions. One is the notion, introduced in this manuscript, of intentional collective stance, which extends the intentional stance [42] to collectives: we may not really know whether the MAS is an intentional collective and what makes it so, but we may still treat it like it was so, e.g., to support reasoning and design activities. For instance, in the crowd example,

```
val crowdingLevel = crowdTracking(p, r, w, thCrowd, thDanger, t)
crowdDispersal(crowdingLevel, riskRange) // dispersal advice
```

the aggregate MAS may be considered as a collective with intentions, i.e., a single distributed entity whose intentions include leaving its internal components free to move at first but also monitoring and ensuring that they do not gather excessively—as a form of (self-)protection.

The second notion is that of joint intentionality [43]. Indeed, collective intentionality requires agents to be jointly directed towards shared activities (plans) or goals. In aggregate computing, the shared goal and the corresponding shared plan to achieve it are reified into an aggregate program (cf. Figure 2c), namely a program that is meant to be played by the whole MAS. Notice that often, like in the crowd example, the individual goal (e.g., moving to a point of interest in the city) may conflict with the collective goal (e.g., moving in the opposite direction to ensure safe dispersal). Therefore, collective autonomy—as the expression of the goals and intentions of an entire collective—potentially reduces individual autonomy, and vice versa (cf. the notion of endogenous or delegated autonomy in the previous section).

Interestingly, an aggregate program does not just embed the collective goals, but also the collective process leading to the selection of collective intentions. For instance, in the crowd example, dispersal is activated once a group of devices determines that the level of danger is sufficiently high (cf. Line 13 in Figure 2c).

As we said, collective autonomy, as a notion, is relational and gradable. The autonomy of a collective can be related to

- the autonomy of the members of the collective—as discussed;
- the environment—namely the extent to which activity depends on environmental situations and events.

Notice that an aggregate MAS can be collectively autonomous even if its overall behavior is highly determined by the deliberation of few individuals—if those individuals have been delegated for decision-making by the collective. Consider the Self-organising Coordination Regions (SCR) pattern [44], which is also exploited in Section 4; a general encoding in ScaFi is as follows:

```
val leaders = S(grain) // sparse-choice: elect leaders at mean distance of grain
val potential = distanceTo(leaders) // gradient field from leaders
val regions = broadcast(potential, mid()) // multi-hop propagation of IDs of leaders
val collectedData = C(potential, membersData()) // collect info towards leaders
val decisions = broadcast(potential, leaderDecision(collectedData)) // propagate leader choices
localAction(decisions)
```

The leader agents (i.e., those for which the `leaders` Boolean field holds `true`) are responsible for making decisions about how the members of the corresponding areas are to behave, but those leaders are elected through a collective process represented by function S [45] (where "S" is a contraction of "sparse-choice", i.e., typically a spatially uniform selection of nodes in a situated network). Function S aims at selecting leaders at a mean distance of `grain` among them. For the pattern to work, further collaboration is needed among the agents to propagate information. Function `distanceTo(s)` is used to compute a self-healing gradient field [46], i.e., a self-stabilising [11] computational field of minimum distances from any node in the system to the node(s) where `s` is `true`, which is also able to correct the individual estimations by reacting to changes of sources, neighbours, and corresponding positions. Such a gradient field can effectively act as a "`potential field`", namely as a kind of force providing a direction and intensity with respect to a reference point, e.g., for moving information or agents [47]. Indeed, function `broadcast(p,v)` is used to implement a multi-hop information stream of the value v at nodes of null potential (i.e., where p is 0) outwards by "ascending" the potential field p. Dually, function `C(p,v)` (for "collection") provides an information stream converging towards nodes at null potential (i.e., where p is 0); sometimes, an operator is provided to specify how the information should be aggregated along the path (e.g., when collecting sets of data, the set union operator may be used to aggregate all the data elements).

We also stress that the collective behavior is not merely the sum of the individual behaviors but the emergent result of repeated individual behaviors involved in a complex network of interactions among related agents (neighbors) and with the environment as well. In other words, an aggregate program provides a schema for self-organizing, self-adaptive behavior that is instantiated once a proper dynamic (defined in terms of a concrete aggregate execution and environmental evolution) is injected over it.

3.4. Summary and Comparison with Related Work

To recap, aggregate programming provides computational mechanisms for supporting various levels of autonomy. Examples include:

- collective autonomy: by cooperative execution of the aggregate program, or in terms of collective structures constraining individual behavior;
- (endogenous) individual autonomy: by calling local functions (as a sort of delegation);
- adjustable autonomy: by using structures or branching mechanisms to regulate the relationship between individuals and collective autonomy, or by controlling the amount of endogenous vs. exogenous autonomy (cf. Table 2).

The discussion in this section is substantiated by the experiments of Section 4 and extended in Section 5 with further considerations on research gaps and potential directions for future investigations.

The proposed approach differs from other works (such as those reviewed in Section 2) in crucial ways. Prominently, it takes a global ("aggregate") perspective on MAS design and programming. Other approaches, such as AgentSpeak(L)/Jason [24], define a MAS by focussing on the behavior of the individual agents, expressed in terms of a number of plans describing how individual goals are to be achieved. However, a Jason program does not directly model collective decision-making or collective action. Note that such perspectives are not alternative but complementary: the support for reasoning available in Jason, based on the BDI architecture, for determining what individual autonomous behavior has to be enacted is not built into aggregate programming. As discussed in

Section 5, the combination of cognitive architectures with collective adaptive systems is a theme still to be investigated. The proposed architecture (cf. Figure 1) is simpler than cognitive architectures: it is inspired by self-organising systems [48], and fosters emergence of collective behaviour rather than reasoning.

Works that also consider MASs by a global perspective include organization-oriented programming approaches such as MOISE [28]. MOISE is a very articulated approach for describing agent organisations (i.e., dynamic groups of agents comprising roles, properties, and interaction protocols) and how agents and organisations interact (e.g., participation to organisations, evolution of organisations, and organisational effects on agents). It is a very rich and flexible model, but it is also quite complex, requiring explicit definitions for the structural, functional, and normative dimensions. By contrast, the aggregate computing approach expresses collective behavior through relatively small scripts (cf. the example in Figure 2c) obtained by composing functions of other collective behavior together. That is, whereas MOISE builds on an explicit representation of organizations, with agents knowing and reasoning about them, aggregate computing favors a more implicit representation of group structures, more typical in the swarm intelligence and self-organisation tradition. As aggregate computing builds on the main abstraction, the computational field, and neighbor-oriented communication, there is low conceptual overhead, with the functional abstraction enabling fine-grained problem decomposition. On the other hand, other approaches of autonomous ensembles programming, such as SCEL [49], do not achieve the same levels of declarativity of aggregate programming languages such as ScaFi. Arguably, the investigation of notions like collective autonomy could take advantage of "simple", compact models which are however able to represent emergent, collective phenomena.

4. Case Study

To showcase the ability of aggregate computing to orchestrate collective behavior comprising both individual and collective autonomy, in this section we present a case study, evaluated by simulation. We conduct the experiments using Alchemist [50], a bio-inspired large-scale multi-agent simulator supporting the ScaFi aggregate programming language [35], which is used to write aggregate programs in our experiments.

Our basic requirements for a simulator include the ability of simulating large-scale networks of mobile devices as well as the ability of defining a dynamics suitable to express the aggregate execution protocol discussed in Section 3.1, which consists of asynchronous rounds of execution with neighborhood-based communication. Therefore, Alchemist represents a first choice as it ships already with a module providing ScaFi support. Moreover, Alchemist is solid and flexible: in the literature, it has been used extensively to simulate multi-agent and aggregate computing systems, in scenarios including crowd simulation [8], drone swarms [34], edge computing [51], and people rescue [36].

The simulations are open-sourced and accessible from the following public repository https://github.com/cric96/mdpi-jsan-2020-simulation, accessed on 1 April 2021.

We consider wildlife monitoring as target application domain, which nowadays starts to take advantages of Internet-of-Things (IoT) infrastructure, unmanned ground (UGVs) or aerial (UAVs) vehicles [52], as well as wearables for animals (such as smart collars) or operators [53]. In this scenario, developers could use MAS programming approaches, such as aggregate computing, to coordinate multiple nodes and agents in performing collective activities such as rescuing animals in danger, geofencing, and detecting/tracking poachers. In these experiments, our emphasis is on the specification of autonomous behavior, as well as the emergent relationship between local and collective autonomy. Bridging with realistic environments and data is left as future work.

4.1. Experiment Setup

The collective goal is to find animals in danger and rescue them. The environment consists of a continuous two-dimensional space with an area of 2500 m × 2500 m. There are three types of nodes:

- Animal: an agent that needs to be rescued if it is in danger. The danger status changes are domain-specific dynamic. In this simulation, the animals change their status randomly at a constant rate (one animal every two seconds). It could be a sort of "smart collar".
- Mobile node: an agent that moves around the world and could have the capability to heal an animal.
- Station: a fixed node that works as a gateway for mobile nodes.

In this experiment, Mobile nodes can perform tasks, namely problems to be solved. Our characterization of this concept is a lightweight version of the definition in [54]: each task has a goal, a set of capabilities needed to accomplish it, and spatial constraints (e.g., on the location of the agent or the task). We do not consider the deadline and priority concepts. The tasks that are created collectively (e.g., by the leader on the basis of data collected from other agents) are called Collective tasks. Local tasks are those crafted by the agents themselves based on local perceptions and intentions. We have identified two types of tasks:

- ExploreAreaTask: leads agents to explore an established portion of the space.
- HealTask: for which a set of agents must rescue a defined animal (and must correspondingly have "rescue capabilities").

At runtime, the program identifies agent roles (healer, explorer, stationary) according to the local behaviour sensed. There are 80 mobile nodes (40 explorers and 40 healers), 20 stations and 100 animals. Only healers can rescue animals. The animals are divided into five independent groups. Each group moves according to the random waypoint logic [55]. An animal in danger needs a variable number of healers near to him to be rescued. The program follows the Self-organising Coordination Regions (SCR) pattern [44]:

1. leader election is made in the stationary nodes (via S);
2. mobile nodes sense animals in danger and send the local information to the leader (via C);
3. the leader chooses what is the animal that needs to be rescued;
4. the leader shares its choice (via broadcast);
5. the slaves act according to the leader choice.

To be more specific, the program first tries to detect nodes' role checking how much each node has moved in a defined time window:

```
val movementWindow = 6 // a domain-specific parameter
val movementThr = 1.0 // another domain-specific parameter
val trajectory = recentValues(movementWindow, currentPosition())
val lastPointInTrajectory = trajectory.reverse.head
val distanceApprox = trajectory.head.distanceTo(lastPointInTrajectory)
mux (distanceApprox <= movementThr) "stationary" else "explorer"
```

After, the collective chooses leaders using block S that is executed only where isStationary is true (using branch). At this point, mobile nodes check if there is an animal in danger in their neighborhood (i.e., they check the status of a special sensor called danger). To do this, we use the operator foldhood(init)(acc)(expr) to aggregate values over neighborhoods with operator acc and initial/null value init; the values that are aggregated are the results of the neighbors' evaluations of expr (where the neighbor-specific parts are defined through operator nbr). Then the information is sent to leaders through the C construct. This block aggregates and moves the data toward a potential field. In this case, the potential field is centered where the leader value is true. Leaders then receive a map specifying the approximate location of animals in danger and their ID. So they peek an animal with a local policy (e.g., choosing the nearest node) and then they create a HealTask that has the goal of directing healers to a chosen animal to rescue it. Besides, leaders share tasks that lead the explorers to stay in the leader's area of influence (i.e., creating an ExploreAreaTask). At this point, tasks are shared across the zones with the broadcast function. Finally, the slaves, who receive the various tasks created, choose which one to execute according to their role and local intentions (e.g., a node may not listen to the collective's decision and act by its own). With ScaFi, this behavior can be synthesized as described in Figure 3b.

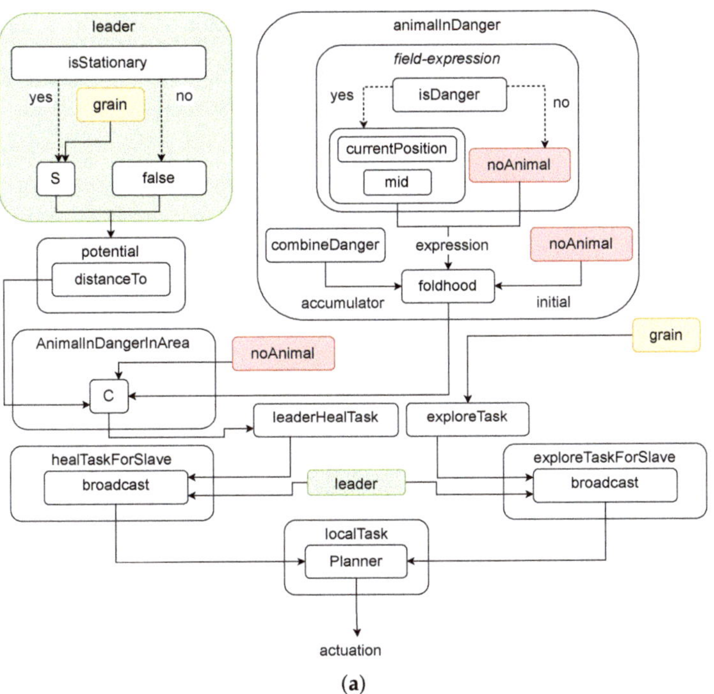

(a)

```scala
val leader = branch(isStationary) { S(grain) } { false } //(1.) leader election
//.. (2.) sense animals in danger ..
val noAnimal : Map[ID, P] = Map.empty
val animalsInDanger = {
  foldhood(noAnimal)(combineDangerMap){
    mux(nbr(isDanger)) {
      Map(nbr(mid() -> currentPosition()))
    } { noAnimal }
  }
}
val potential = distanceTo(leader)
//(2.) send information to the leader
val animalDangerInArea = C(potential, combineDangerMap, animalsInDanger, noAnimal)
//(3.) leader chooses an animal to rescue
val leaderHealTask = animalDangerInArea.toSeq.sortBy {
  case (_, p) => p.distance(currentPosition()) //choosing policy
}.headOption.map {
  case (id, p) => HealTask(mid(), id, p)
}
val exploreArea = ExploreTask(mid(), currentPosition(), grain) //for explorers
//(4.) and share its choice via broadcast
val healTaskForSlave = broadcast(leader, leaderHealTask)
val exploreAreaToSlave = broadcast(leader, exploreArea)
//(5.) slave choose the task according its role, intentions and leader choice
val collectiveTasks = Seq(healTaskForSlave, exploreAreaToSlave)
val localTask = Planner.eval(collectiveTasks) //the task choice is encapsulated here..
val actuation = localTask.call(this) // produces data in order to achieve the task chosen
```

(b)

Figure 3. Wildlife monitoring and rescue example: architecture and program. (**a**) Diagram corresponding to the code below. This figure uses the same notation of Figure 2b. In addition, we use a same colour to denote multiple references to the same functional block. (**b**) The ScaFi code snippet for the wildlife program behaviour.

Next, we briefly link different behaviors with different levels/types of autonomy. From the local perspective, each agent: (i) moves around the environment; (ii) executes tasks. Each behavior has an endogenous and exogenous aspect: with this program, we can specify how agents should behave (endogenous) but we have not total control over them because they maintain a certain level of exogenous autonomy. For example, when an agent receives the task to rescue an animal, he might decide to rescue another animal because he finds another one closer or because in that direction, there are obstacles. Moreover, the endogenous/exogenous autonomy level depends on the agent type. Indeed, we cannot control animal movements via the smart collar, so they maintain a total level of exogenous autonomy with respect to the program. From a collective perspective, the MAS performs: (i) multileader election, (ii) animal dangers sensing, and (iii) task selection to reduce the animal in danger in zones. With ScaFi the overall behaviour is intentionally described as:

```
val targets = animalsInDanger()
branch(isAnimal){doNothing} {rescue(targets)}
```

where `rescue(targets)` performs the code above and `animalsInDager` executes the neighbourhood perception danger sensor.

What we want to enforce in this experiment is that individual autonomy influences collective autonomy. For example, the leader election happens on stationary nodes. However, the node stationary rule depends on the local agent activity (mainly how they move), which is an exogenous behavior. So, what we expect is that most the system is locally autonomous less the collective autonomy influences the overall behavior.

4.2. Performance Metrics and Parameters

In simulation evaluation, we consider whether functional and non-functional aspects. The main functional metric is the rescue count, which describes how many animals are rescued during the simulation. Another condition that will be evaluated is the number of leaders without healers necessary to heal an animal. A non-functional value verified is the average distance to the leader. Ideally, agents might be arranged to cover a zone uniformly.

We also introduce a metric that represents the general performance of a simulation run. Given a period t expressed in seconds, we sample the experiment at each second. For each sample, we evaluate two parameters, *events* and *healed count*. Given a sample t, events measure how many animals turn on danger during the simulation until the t. Indeed, the healed count tells how many animals were rescued by the collective. Hypothetically, if the system is completely reactive, healed count and events should have the same value. Hence, the greater is the distance between these two values, the worse system perform. So, as comprehensive system performance, we decide to use Root Mean Squared Error (RMSE):

$$Error = \sqrt{\frac{1}{t} * \sum_{i=1}^{t} (events_i - healedcount_i)^2}$$

What we expect from these simulations is that selfish settings bring worse performance. Indeed, here we need a collective choice to rescue an animal. We test the application varying these parameters:

- p: is the probability to follow the collective choice, $1 - p$ is the probability to act selfishly; the bigger is p the lesser the agent is autonomous with respect to the collective goals.
- healer count: the nodes needed to rescue an animal. A higher value of healer count needs greater control on local agent behaviour in order to accomplish the collective task

4.3. Results and Discussion

We now present the key results produced by the different simulation runs we conducted. In Figure 4 there is a graphical result obtained in Alchemist. We had varied p by four values (0, 0.25, 0.5, 0.75 and 1) and healer count by four values (2, 4, 6, 8). For each p and healer count combination, 50 simulation runs are performed. The data reported in Figure 5 contains the average behavior of those simulations. In Figure 6, the average error is reported for each combination of healer count and p. Each run has lasts 300 s. Furthermore, after 100 s, no other animals turn their status in danger. It helps us to see how what system is faster to return into a stable configuration.

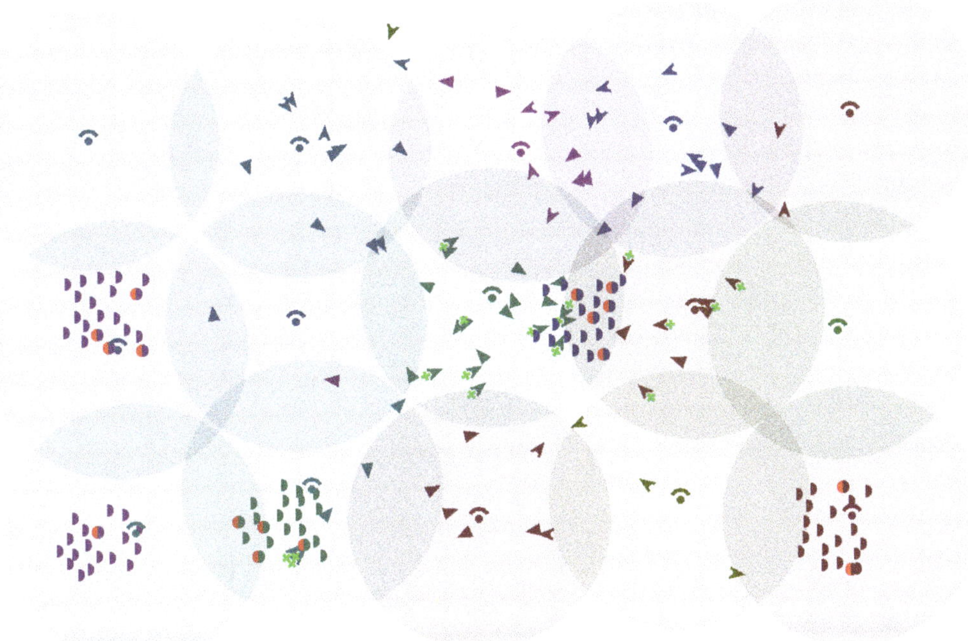

Figure 4. Wildlife monitoring simulation screenshot. The "wi-fi" like symbol represents a Station node. Each Station has an influence area (displayed with a colored circle). Mobile nodes are represented with a triangle. Healers have a half-moon on the shorter side. When healers have an animal target, a green cross appears. Animals are drawn as a half-circle. Each group has a uniform color. When they turn in danger, a half red circle comes out.

In general, the results confirm our thesis: the more agents act autonomously, the worse the system behaves. However, the performance depends upon the application domain. Indeed, when healer count is little, the overall behaviors aren't so different. Because even if agents do not collaborate, there is a high probability that two Mobile nodes are near to the same animal in danger. We see the benefits of collective choice with higher healer count values (Figure 6). For example, when healer count is 8 the system at $p = 1$ performs better than all other configurations. We want to emphasize how the difference w.r.t. the other p values grows as healer count increases. The difference instead becomes marked when healer count is 8, since the task has a greater need for collective choice and even small local selfish choices lead to worse performance. Outside of the functional aspects, we can clearly see that the higher the p value the more the system follows joint intentionality: agents arrange themselves equally in each area (having a higher average) and they cover the zones uniformly. This was an expected result, but it also makes us understand that even if the system loses some level of collectivity, it still manages to perform well (for example, if 25% of the time agents act independently, namely when p is 0.75, the overall performance is practically equal to when the agents always follow the collective choice). In general, however, we can observe that even if the system control is not total, the emerging behavior is comparable with others. It is because all agents must adapt to the aggregate protocol, losing some of their autonomy.

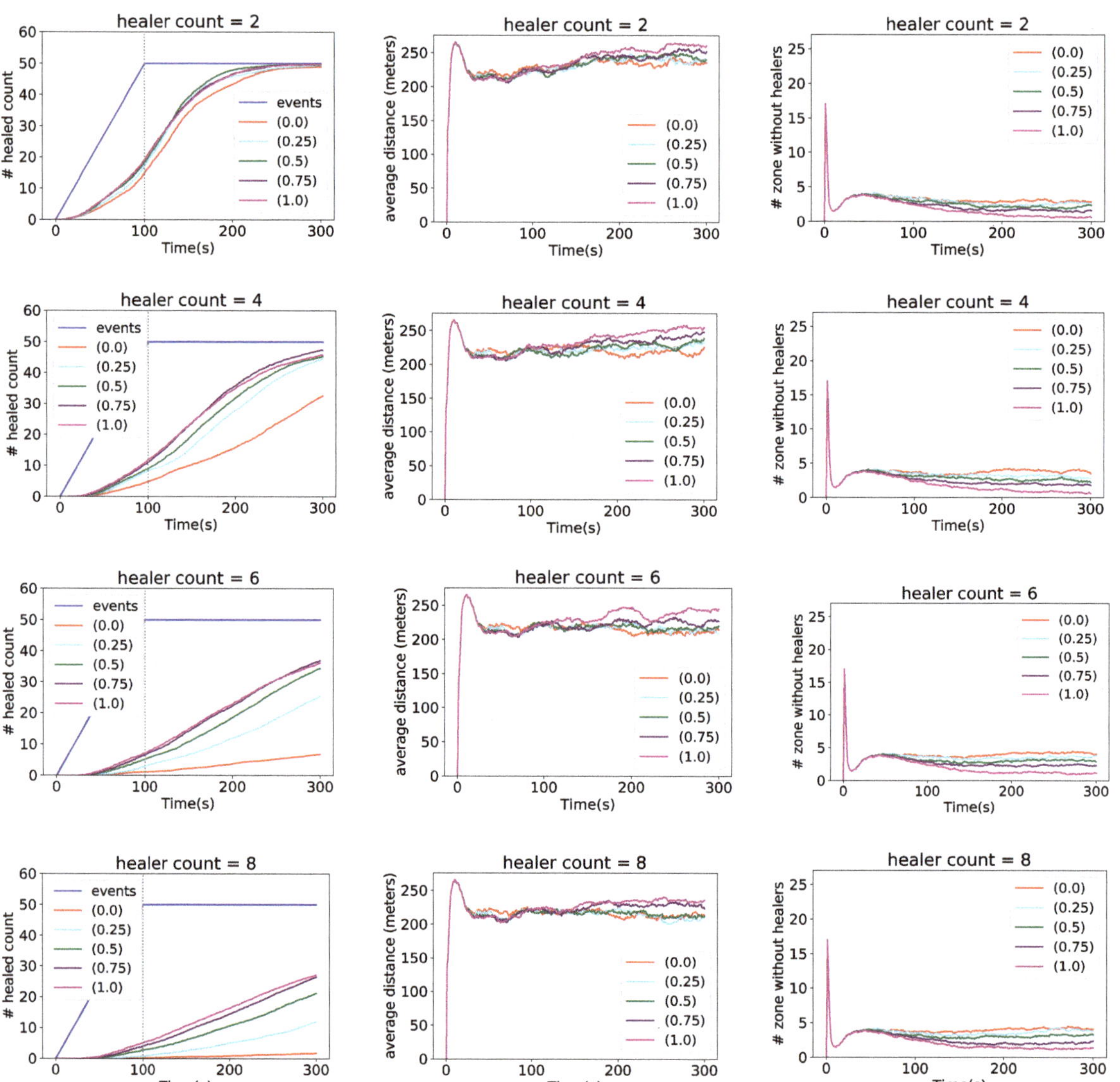

Figure 5. Experiment results. In each plot, the color identifies the p parameter. The first column shows how many animals are rescued during the simulation. The plots in the second column show the average distance from the leader. The plots in the last column point out how many zones have not enough healers to rescue an animal.

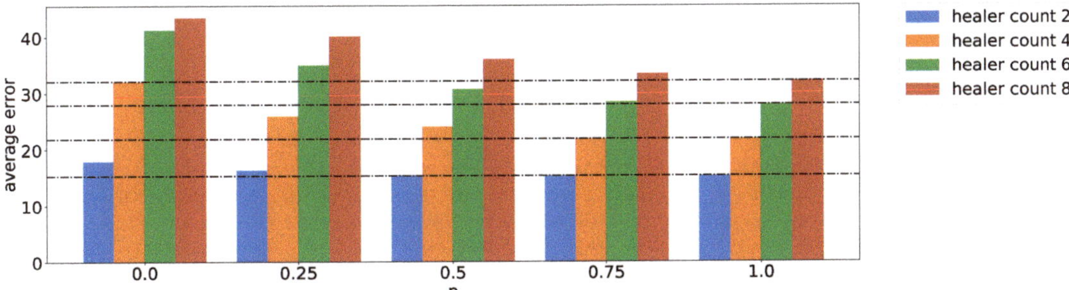

Figure 6. Average error (RMSE) for each combination of *p* and healer count. Horizontal lines mark the performance of *p* = 1.0.

4.4. Final Remarks

In this experiment, we deliberately did not consider the following aspects (even if they could affect the system performance), which are abstracted away: how the system detects animals in danger; the physical/hardware details of system nodes; how specific capabilities provided by the different node roles are implemented as concrete actions. Defining when an animal is in danger is a complex task per se and is strongly domain-dependent. In [56], the authors use UAVs to verify the presence of poachers. Concerning geofencing, in [57] the authors use a collar equipped with GPS to verify when animals escape from a boundary and mark them in danger. Animal immobility is another alarm signal used in [58]. Roles are introduced in this paper, though they are not present in other works (to the best of our knowledge), as they are functional to divide the work between nodes. In general, we expect that nodes could either be humans (cf., healers), UAVs (cf., explorers), or installed gateways. In the survey [59], some interesting work such as [60] does not require human intervention to rescue animals (cf., healer role). Finally, we want to underline that our approach is on a higher level of abstraction, focussing on coordination and execution of collective autonomous behavior. Environmental aspects are perceived through sensors that can be as simple (e.g., temperature sensor) or as complex as desired (e.g., a scheduler to choose tasks).

5. Research Roadmap

In the following, we identify a roadmap of two research directions to unveil the full potential of aggregate programming for collective autonomy: they comprise expressing collective autonomous behavior (Section 5.1) and achieving reliable collective autonomy (Section 5.2).

5.1. Expressing Collective Autonomous Behaviour

Addressing problems at a suitable level of abstraction is key in modeling and programming. In this section, we cover two interesting research directions related to specifying collective autonomy, and point out a few starting points in aggregate programming research.

5.1.1. Cognitive Collectives

Notions like distributed cognition and group minds are often leveraged in sociology and cognitive sciences to explain collective processes. In some works, such as in [61], a dynamical systems approach is used to model and understand the emergence of collective behavior and collective forms of consciousness. There, formal arguments are provided to explain how groups exert downward causation on their components. In the aggregate approach, groups or collectives are the target of programming, and the downward causation is a straightforward consequence of being part of the aggregate. However, collective cognitive states are not explicitly represented, but rather mixed in the aggregate programs, which can be seen as shared plans. Indeed, it would be interesting to investigate, along the lines of [36], whether more explicit representations of collective cognition might help to drive group and individual behavior.

5.1.2. Collective Autonomy and Structural Organisation

One of the first works analyzing the relationship between organizational structure and autonomy is by Schillo [62]. Schillo proposes a framework based on the notion of delegation, draws a link between self-organization and adjustable autonomy, and defines a spectrum of seven organizational forms of increasing coupling between agents in a MAS: single-agent system, task delegation, virtual enterprise, cooperation, strategic network, group, cooperation. In [63], Horling et al. provide a survey of MAS organizational paradigms. They identify ten kinds of organizational structures: adhocracies, hierarchies, holarchies, coalitions, teams, congregations, societies, federations, markets, and matrix-like organizations. We believe that more work is needed for better understanding and specifying the inter-dependencies between structures and collective autonomous behavior.

For instance, approaches to MAS programming based on the notion of a team (i.e., a sub-collective) have been proposed in the past [64–66]. In the context of aggregate computing, the aggregate process abstraction—modeling a concurrent, dynamically scoped aggregate computation—has been recently introduced to extend the practical expressiveness of the paradigm. Aggregate processes, by defining concurrent activities with a dynamic and possibly overlapping scope, would support the specification of the autonomous behavior of multiple collectives. In this respect, it would be interesting to investigate the relationships between the autonomies of different collectives, as well as the interaction between the autonomies of a collective and its sub-collectives. Approaches inspired by holonic MASs [48], such as SARL [27], may prove useful or insights when considering multiple levels of autonomy and hierarchical organizations.

5.2. Reliable Collective Autonomy

Related to the ability of expressing collective autonomous behavior is the extent to which specifications lead to properties for dependable system operation that can be promoted, verified, and formally guaranteed.

5.2.1. Safety and Guarantees

Currently, few results are available for programming reliable collective adaptive behaviors. In the context of aggregate programming, major formal results include self-stabilisation [11] and eventual consistency to device distribution [39]. These results, however, are typically valid for restricted fragments of the field calculus. Moreover, self-stabilization does not say much about the speed of convergence, nor the ability of an algorithm to withstand continuous change. Therefore, typical validation approaches also include simulation [67] as a key step. Distributed runtime verification may also prove useful [68]. In general, more work is needed towards methods, both formal and lightweight, to verify the correctness and provide guarantees about global results in a certain range of environments and conditions.

5.2.2. Norms and Trust

Among the notions studied in literature to promote good cooperation between agents there are norms and trust. Through norms, and corresponding mechanisms for enforcing them (such as sanctions, rewards, and institutions), it is possible to regulate individual and then collective activity for social benefits. The problem is to make the MAS determine what deviant behavior is, detect it, and take corresponding countermeasures. Norms may be determined collectively, through processes of agreement or conflict resolution [69]. However, few results are available on conflicts among multiple norms and steering of their emergent effects at the collective level [69].

Trust can also be used as a way to reduce cooperation inefficiency and issues. An excess of individual autonomy may lead to deviance. In large-scale MASs, even few cases of deviance may result in serious emergent effects at the collective level [70]. Preliminary work in aggregate programming research has been carried out to exclude non-trusted devices from the collective computation [71], leveraging a notion of trust field mapping any agent of the MAS to a trust score that is computed collectively. We believe further work is required on these topics to promote reliable (collective) autonomy.

5.3. Applications

Multi-agent systems and technologies span various application domains [72] including e-commerce [73], health care [74], logistics [75], robotics [76], manufacturing [77] and energy [78,79]: they are potential application areas for the approach to aggregate programming presented in this paper. The survey by Müller et al. [72] provides an overview of the impact of deployed MAS-based applications, showing that MAS technology has already been successful in various sectors. It is expected that MASs would be increasingly significant in the future, as more and more devices get deployed in our environments (cf. IoT, CPS, and related trends) and visions like autonomic computing continue to develop, fostering a pervasive embedding of computational autonomy at various levels. This is also plausible for collective autonomy, as applications involving (cyber–physical) collectives of (variously autonomous) actors seeking global goals emerge—cf. applications in smart city [80], swarm robotics [81], mobile social crowdsensing [82,83], and smart infrastructures contexts [51,84]. In particular, programmable approaches to collective autonomy could contribute to research and applications of computational collective intelligence [85], namely the field studying groups and their ability to implement effective decision-making, coordinated action, and cooperative problem-solving. However, we also remark that approaches to designing and programming MASs at the collective level should not be considered as omni-comprehensive approaches that deal with every aspect of a system; instead, they represent a tool that can be used to address certain problems (i.e., promoting global behavior and properties) at a suitable abstraction level. In this sense, a further challenge to be addressed in the future is the integration of collective-level approaches with individual-level approaches and traditional paradigms, promoting the vision of system development through integration of multiple perspectives and viewpoints [86].

6. Conclusions

In this paper, we consider the problem of programming the collective autonomous behavior of multi-agent systems. We consider the aggregate programming paradigm, a framework founded on a calculus of computational fields originally introduced to express the coordination and self-organisation logic of spatially situated systems. We analyze the support provided by aggregate computing by a MAS perspective, and, as a contribution, (i) interpret its execution protocol as an agent control architecture; and (ii) analyze aggregate programs by the point of view of individual and collective autonomy. Finally, we provide some simulation-based experiments, to show how the framework supports analysis of autonomy-related aspects, and discuss research gaps, pointing out opportunities for new research.

The various goals described in the introduction have been addressed as follows. The literature review in Section 2 provides a variegated view of autonomy in software engineering and MAS, with emphasis on actionable notions and programming. This helped us to position the contribution of Section 3 on aggregate computing, where multiple autonomy-related notions are supported (also enabling adjustable autonomy) but implicit, and had never been unveiled in previous publications. Explicit mechanisms and extensions for (adjustable) autonomy could be considered in the future. Moreover, the contribution of the view of the aggregate execution model as an agent control architecture represents a step towards comparison and integration with other architectures—which is left as interesting future work. The case study of Section 4 shows how the discussed framework enables functional specification of MASs exhibiting a form of collective autonomy as well as parameterized behaviors where individual and collective autonomy can be adjusted off-line. Qualitatively, it shows that the approach is feasible and may scale with complexity. Self-organizing algorithms and patterns for on-line autonomy adjustment in aggregate systems as well as more quantitative analysis of trade-offs can be considered in future research. Finally, the discussion in Section 5 highlights significant research directions and application domains which complement the above discussion.

We argue that it is important to directly address the collective dimension of MASs, rather than programming individual agents and then verifying that local behaviors lead to the intended, but not explicitly captured, global behavior. It is a matter of abstraction and addressing concerns from a proper perspective. As discussed, specifications of collective autonomous behavior, such as aggregate programs written in ScaFi, provide natural support for adjustable autonomy, and this could pave the way to the integration with traditional agent control architectures. However, more work is needed to ensure that the collective specifications result in acceptable emergent behaviors, and hence in reliable forms of collective autonomy.

Author Contributions: Conceptualization, R.C., G.A. and M.V.; methodology, R.C. and M.V.; software, R.C., G.A. and M.V.; validation, R.C. and G.A.; formal analysis, G.A.; investigation, R.C. and G.A.; resources, R.C. and M.V.; data curation, G.A.; writing—original draft preparation, R.C. and G.A.; writing—review and editing, R.C., G.A. and M.V.; visualization, R.C. and G.A.; supervision, M.V.; project administration, R.C. and M.V. All authors have read and agreed to the published version of the manuscript.

Funding: This research received no external funding.

Data Availability Statement: Refer to https://github.com/cric96/mdpi-jsan-2020-simulation, accessed on 1 March 2021.

Conflicts of Interest: The authors declare no conflict of interest.

References

1. Kephart, J.O.; Chess, D.M. The Vision of Autonomic Computing. *Computer* **2003**, *36*, 41–50. [CrossRef]
2. de Lemos, R.; Giese, H.; Müller, H.A.; Shaw, M.; Andersson, J.; Litoiu, M.; Schmerl, B.R.; Tamura, G.; Villegas, N.M.; Vogel, T.; et al. *Software Engineering for Self-Adaptive Systems: A Second Research Roadmap*; Springer: Berlin/Heidelberg, Germany, 2010; Volume 7475, pp. 1–32.
3. Wooldridge, M.J. *An Introduction to MultiAgent Systems*, 2nd ed.; Wiley: Hoboken, NJ, USA, 2009.
4. Ferber, J. *Multi-Agent Systems—An Introduction to Distributed Artificial Intelligence*; Addison-Wesley-Longman: Boston, MA, USA, 1999.
5. Weyns, D.; Omicini, A.; Odell, J. Environment as a first class abstraction in multiagent systems. *Auton. Agents Multi Agent Syst.* **2007**, *14*, 5–30. [CrossRef]
6. Beal, J.; Dulman, S.; Usbeck, K.; Viroli, M.; Correll, N. Organizing the Aggregate: Languages for Spatial Computing. *arXiv* **2012**, arXiv:1202.5509.
7. Viroli, M.; Beal, J.; Damiani, F.; Audrito, G.; Casadei, R.; Pianini, D. From distributed coordination to field calculus and aggregate computing. *J. Log. Algebraic Methods Program.* **2019**, 100486. [CrossRef]
8. Beal, J.; Pianini, D.; Viroli, M. Aggregate Programming for the Internet of Things. *Computer* **2015**, *48*, 22–30. [CrossRef]
9. Audrito, G.; Viroli, M.; Damiani, F.; Pianini, D.; Beal, J. A Higher-Order Calculus of Computational Fields. *ACM Trans. Comput. Log.* **2019**, *20*, 5:1–5:55. [CrossRef]
10. Beal, J.; Viroli, M. Building Blocks for Aggregate Programming of Self-Organising Applications. In Proceedings of the Eighth IEEE International Conference on Self-Adaptive and Self-Organizing Systems Workshops, SASOW 2014, London, UK, 8–12 September 2014; pp. 8–13. [CrossRef]
11. Viroli, M.; Audrito, G.; Beal, J.; Damiani, F.; Pianini, D. Engineering Resilient Collective Adaptive Systems by Self-Stabilisation. *ACM Trans. Model. Comput. Simul.* **2018**, *28*, 16:1–16:28. [CrossRef]
12. Malone, T.W.; Crowston, K. The interdisciplinary study of coordination. *ACM Comput. Surv.* **1994**, *26*, 87–119. [CrossRef]
13. Odell, J. Objects and Agents Compared. *J. Object Technol.* **2002**, *1*, 41–53. [CrossRef]
14. Franklin, S.; Graesser, A. *Is It an Agent, or just a Program? A Taxonomy for Autonomous Agents*; International Workshop on Agent Theories, Architectures, and Languages; Springer: Berlin/Heidelberg, Germany, 1996; pp. 21–35.
15. Castelfranchi, C.; Falcone, R. Founding Autonomy: The Dialectics Between (Social) Environment and Agent's Architecture and Powers. In *International Workshop on Computational Autonomy*; Springer: Berlin/Heidelberg, Germany, 2003; Volume 2969, pp. 40–54.

16. Castelfranchi, C. Guarantees for autonomy in cognitive agent architecture. In *International Workshop on Agent Theories, Architectures, and Languages*; Springer: Berlin/Heidelberg, Germany, 1994; pp. 56–70.

17. De Silva, L.; Meneguzzi, F.; Logan, B. BDI Agent Architectures: A Survey. In Proceedings of the International Joint Conferences on Artificial Intelligence, Yokohama, Japan, 11–17 July 2020.

18. Sekiyama, K.; Fukuda, T. Dissipative structure network for collective autonomy: Spatial decomposition of robotic group. In *Distributed Autonomous Robotic Systems 2*; Springer: Berlin/Heidelberg, Germany, 1996; pp. 221–232.

19. Bottazzi, E.; Catenacci, C.; Gangemi, A.; Lehmann, J. From collective intentionality to intentional collectives: An ontological perspective. *Cogn. Syst. Res.* **2006**, *7*, 192–208. [CrossRef]

20. Ray, P.; O'Rourke, M.; Edwards, D. Using collective intentionality to model fleets of autonomous underwater vehicles. In Proceedings of the OCEANS 2009, Biloxi, MS, USA, 26–29 October 2009; pp. 1–7.

21. Conte, R.; Turrini, P. Argyll-Feet giants: A cognitive analysis of collective autonomy. *Cogn. Syst. Res.* **2006**, *7*, 209–219. [CrossRef]

22. Liu, J.; Jin, X.; Tsui, K.C. Autonomy-oriented computing (AOC): Formulating computational systems with autonomous components. *IEEE Trans. Syst. Man Cybern. Part A* **2005**, *35*, 879–902. [CrossRef]

23. Mao, X.; Wang, Q.; Yang, S. A survey of agent-oriented programming from software engineering perspective. *Web Intell.* **2017**, *15*, 143–163. [CrossRef]

24. Bordini, R.H.; Hübner, J.F.; Wooldridge, M. *Programming Multi-Agent Systems in AgentSpeak Using Jason*; John Wiley & Sons: Hoboken, NJ, USA, 2007; Volume 8.

25. Finin, T.W.; Fritzson, R.; McKay, D.P.; McEntire, R. KQML As An Agent Communication Language. In Proceedings of the Third International Conference on Information and Knowledge Management (CIKM'94), Gaithersburg, MD, USA, 29 November– 2 December 1994; ACM: New York City, NY, USA 1994; pp. 456–463. [CrossRef]

26. Ricci, A.; Piunti, M.; Viroli, M.; Omicini, A. Environment Programming in CArtAgO. In *Multi-Agent Programming, Languages, Tools and Applications*; Bordini, R.H., Dastani, M., Dix, J., Seghrouchni, A.E.F., Eds.; Springer: Berlin/Heidelberg, Germany, 2009; pp. 259–288. [CrossRef]

27. Rodriguez, S.; Gaud, N.; Galland, S. SARL: A General-Purpose Agent-Oriented Programming Language. In Proceedings of the 2014 IEEE/WIC/ACM International Joint Conferences on Web Intelligence (WI) and Intelligent Agent Technologies (IAT), Warsaw, Poland, 11–14 August 2014; pp. 103–110. [CrossRef]

28. Hübner, J.F.; Sichman, J.S.; Boissier, O. Developing organised multiagent systems using the MOISE$^+$ model: Programming issues at the system and agent levels. *Int. J. Agent Oriented Softw. Eng.* **2007**, *1*, 370–395. [CrossRef]

29. Cardoso, R.C.; Ferrando, A. A Review of Agent-Based Programming for Multi-Agent Systems. *Computers* **2021**, *10*, 16. [CrossRef]

30. DeHon, A.; Giavitto, J.; Gruau, F. (Eds.) *Computing Media and Languages for Space-Oriented Computation*; Schloss Dagstuhl-Leibniz-Zentrum für Informatik: Schloss Dagstuhl, Germany, 2007; Volume 06361.

31. Mamei, M.; Zambonelli, F. *Field-Based Coordination for Pervasive Multiagent Systems*; Springer Series on Agent Technology; Springer: Berlin/Heidelberg, Germany, 2006. [CrossRef]

32. Parunak, H.V.D.; Brueckner, S.; Sauter, J.A. Digital pheromone mechanisms for coordination of unmanned vehicles. In Proceedings of the First International Joint Conference on Autonomous Agents & Multiagent Systems, AAMAS 2002, Bologna, Italy, 15–19 July 2002; pp. 449–450. [CrossRef]

33. Casadei, R.; Viroli, M.; Audrito, G.; Damiani, F. FScaFi: A Core Calculus for Collective Adaptive Systems Programming. Leveraging Applications of Formal Methods, Verification and Validation: Engineering Principles. In Proceedings of the 9th International Symposium on Leveraging Applications of Formal Methods, ISoLA 2020, Rhodes, Greece, 20–30 October 2020; Lecture Notes in Computer Science; Margaria, T., Steffen, B., Eds.; Springer: Berlin/Heidelberg, Germany, 2020; Volume 12477, pp. 344–360. [CrossRef]

34. Casadei, R.; Viroli, M.; Audrito, G.; Pianini, D.; Damiani, F. Engineering collective intelligence at the edge with aggregate processes. *Eng. Appl. Artif. Intell.* **2021**, *97*, 104081. [CrossRef]

35. Casadei, R.; Pianini, D.; Viroli, M. Simulating large-scale aggregate MASs with Alchemist and Scala. In Proceedings of the 2016 Federated Conference on Computer Science and Information Systems (FedCSIS), Gdansk, Poland, 11–14 September 2016; pp. 1495–1504.

36. Viroli, M.; Pianini, D.; Ricci, A.; Croatti, A. Aggregate plans for multiagent systems. *Int. J. Agent Oriented Softw. Eng.* **2017**, *5*, 336–365. [CrossRef]

37. Casadei, R.; Pianini, D.; Placuzzi, A.; Viroli, M.; Weyns, D. Pulverization in Cyber-Physical Systems: Engineering the Self-Organizing Logic Separated from Deployment. *Future Internet* **2020**, *12*, 203. [CrossRef]

38. Pianini, D.; Casadei, R.; Viroli, M.; Mariani, S.; Zambonelli, F. Time-Fluid Field-Based Coordination through Programmable Distributed Schedulers. *arXiv* **2020**, arXiv:2012.13806.

39. Beal, J.; Viroli, M.; Pianini, D.; Damiani, F. Self-Adaptation to Device Distribution in the Internet of Things. *ACM Trans. Auton. Adapt. Syst.* **2017**, *12*, 12:1–12:29. [CrossRef]

40. Hollander, C.D.; Wu, A.S. The Current State of Normative Agent-Based Systems. *J. Artif. Soc. Soc. Simul.* **2011**, *14*. [CrossRef]

41. Mostafa, S.A.; Ahmad, M.S.; Mustapha, A. Adjustable autonomy: A systematic literature review. *Artif. Intell. Rev.* **2019**, *51*, 149–186. [CrossRef]

42. Dennett, D.C. *The Intentional Stance*; MIT Press: Cambridge, MA, USA, 1989.

43. Huebner, B. *Macrocognition: A Theory of Distributed Minds and Collective Intentionality*; Oxford University Press: Oxford, UK, 2014.

44. Casadei, R.; Pianini, D.; Viroli, M.; Natali, A. Self-organising Coordination Regions: A Pattern for Edge Computing. In Proceedings of the International Conference on Coordination Languages and Models, Kongens Lyngby, Denmark, 17–21 June 2019; Springer: Cham, Switzerland, 2019; pp. 182–199.

45. Mo, Y.; Beal, J.; Dasgupta, S. An Aggregate Computing Approach to Self-Stabilizing Leader Election. In Proceedings of the 2018 IEEE 3rd International Workshops on Foundations and Applications of Self* Systems (FAS*W), Trento, Italy, 3–7 September 2018; pp. 112–117. [CrossRef]

46. Audrito, G.; Casadei, R.; Damiani, F.; Viroli, M. Compositional Blocks for Optimal Self-Healing Gradients. In Proceedings of the 2017 IEEE 11th International Conference on Self-Adaptive and Self-Organizing Systems (SASO), Tucson, AZ, USA, 18–22 September 2017.

47. Wolf, T.D.; Holvoet, T. Designing Self-Organising Emergent Systems based on Information Flows and Feedback-loops. In Proceedings of the First International Conference on Self-Adaptive and Self-Organizing Systems, SASO 2007, Boston, MA, USA, 9–11 July 2007; pp. 295–298. [CrossRef]

48. Rodriguez, S.; Hilaire, V.; Gaud, N.; Galland, S.; Koukam, A. Holonic Multi-Agent Systems. In *Self-organising Software—From Natural to Artificial Adaptation*; Serugendo, G.D.M., Gleizes, M., Karageorgos, A., Eds.; Natural Computing Series; Springer: Berlin/Heidelberg, Germany, 2011; pp. 251–279. [CrossRef]

49. Nicola, R.D.; Loreti, M.; Pugliese, R.; Tiezzi, F. A Formal Approach to Autonomic Systems Programming: The SCEL Language. *ACM Trans. Auton. Adapt. Syst.* **2014**, *9*, 7:1–7:29. [CrossRef]

50. Pianini, D.; Montagna, S.; Viroli, M. Chemical-oriented simulation of computational systems with ALCHEMIST. *J. Simulation* **2013**, *7*, 202–215. [CrossRef]

51. Casadei, R.; Viroli, M. Coordinating Computation at the Edge: A Decentralized, Self-Organizing, Spatial Approach. In Proceedings of the 2019 Fourth International Conference on Fog and Mobile Edge Computing (FMEC), Rome, Italy, 10–13 June 2019; pp. 60–67. [CrossRef]

52. Gonzalez, L.F.; Montes, G.A.; Puig, E.; Johnson, S.; Mengersen, K.L.; Gaston, K.J. Unmanned Aerial Vehicles (UAVs) and Artificial Intelligence Revolutionizing Wildlife Monitoring and Conservation. *Sensors* **2016**, *16*, 97. [CrossRef] [PubMed]

53. Ayele, E.D.; Meratnia, N.; Havinga, P.J.M. Towards a New Opportunistic IoT Network Architecture for Wildlife Monitoring System. In Proceedings of the 9th IFIP International Conference on New Technologies, Mobility and Security, NTMS 2018, Paris, France, 26–28 February 2018; pp. 1–5. [CrossRef]

54. Fatima, S.S.; Wooldridge, M.J. Adaptive task resources allocation in multi-agent systems. In Proceedings of the Fifth International Conference on Autonomous Agents, AGENTS 2001, Montreal, QC, Canada, 28 May–1 June 2001; André, E., Sen, S., Frasson, C., Müller, J.P., Eds.; ACM: New York City, NY, USA, 2001; pp. 537–544. [CrossRef]

55. Mao, S. Chapter 8 - Fundamentals of communication networks. In *Cognitive Radio Communications and Networks*; Wyglinski, A.M., Nekovee, M., Hou, Y.T., Eds.; Academic Press: Oxford, UK, 2010; pp. 201–234. [CrossRef]

56. Mulero-Pázmány, M.; Stolper, R.; van Essen, L.D.; Negro, J.J.; Sassen, T. Remotely Piloted Aircraft Systems as a Rhinoceros Anti-Poaching Tool in Africa. *PLoS ONE* **2014**, *9*, e83873. [CrossRef] [PubMed]

57. Marini, D.; Llewellyn, R.; Belson, S.; Lee, C. Controlling Within-Field Sheep Movement Using Virtual Fencing. *Animals* **2018**, *8*, 31. [CrossRef] [PubMed]

58. Wall, J.; Wittemyer, G.; Klinkenberg, B.; Douglas-Hamilton, I. Novel opportunities for wildlife conservation and research with real-time monitoring. *Ecol. Appl.* **2014**, *24*, 593–601. [CrossRef]

59. López, J.J.; Mulero-Pázmány, M. Drones for Conservation in Protected Areas: Present and Future. *Drones* **2019**, *3*, 10. [CrossRef]

60. Hahn, N.; Mwakatobe, A.; Konuche, J.; de Souza, N.; Keyyu, J.; Goss, M.; Chang'a, A.; Palminteri, S.; Dinerstein, E.; Olson, D. Unmanned aerial vehicles mitigate human–elephant conflict on the borders of Tanzanian Parks: A case study. *Oryx* **2016**, *51*, 513–516. [CrossRef]

61. Palermos, S.O. The Dynamics of Group Cognition. *Minds Mach.* **2016**, *26*, 409–440. [CrossRef]

62. Schillo, M. Self-organization and adjustable autonomy: Two sides of the same coin? *Connect. Sci.* **2002**, *14*, 345–359. [CrossRef]

63. Horling, B.; Lesser, V.R. A survey of multi-agent organizational paradigms. *Knowl. Eng. Rev.* **2004**, *19*, 281–316. [CrossRef]

64. Pynadath, D.V.; Tambe, M.; Chauvat, N.; Cavedon, L. Toward Team-Oriented Programming. In Proceedings of the Intelligent Agents VI, Agent Theories, Architectures, and Languages (ATAL), 6th International Workshop, ATAL '99, Orlando, FL, USA,15–17 July 1999; Jennings, N.R., Lespérance, Y., Eds.; Lecture Notes in Computer Science; Springer: Berlin/Heidelberg, Germany, 1999; Volume 1757, pp. 233–247. [CrossRef]

65. Jarvis, J.; Rönnquist, R.; McFarlane, D.C.; Jain, L.C. A team-based holonic approach to robotic assembly cell control. *J. Netw. Comput. Appl.* **2006**, *29*, 160–176. [CrossRef]

66. Koutsoubelias, M.; Lalis, S. TeCoLa: A Programming Framework for Dynamic and Heterogeneous Robotic Teams. In Proceedings of the 13th International Conference on Mobile and Ubiquitous Systems: Computing, Networking and Services, MobiQuitous 2016, Hiroshima, Japan, 28 November–1 December 2016; Hara, T., Shigeno, H., Eds.; ACM: New York City, NY, USA, 2016; pp. 115–124. [CrossRef]

67. Mittal, S.; Risco-Martin, J.L. Simulation-based complex adaptive systems. In *Guide to Simulation-Based Disciplines*; Springer: Berlin/Heidelberg, Germany, 2017; pp. 127–150.

68. Audrito, G.; Casadei, R.; Damiani, F.; Stolz, V.; Viroli, M. Adaptive distributed monitors of spatial properties for cyber-physical systems. *J. Syst. Softw.* **2021**, *175*, 110908. [CrossRef]

69. dos Santos, J.S.; de Oliveira Zahn, J.; Silvestre, E.A.; da Silva, V.T.; Vasconcelos, W.W. Detection and resolution of normative conflicts in multi-agent systems: A literature survey. *Auton. Agents Multi Agent Syst.* **2017**, *31*, 1236–1282. [CrossRef]
70. Aldini, A. Design and Verification of Trusted Collective Adaptive Systems. *ACM Trans. Model. Comput. Simul.* **2018**, *28*, 9:1–9:27. [CrossRef]
71. Casadei, R.; Aldini, A.; Viroli, M. Towards attack-resistant Aggregate Computing using trust mechanisms. *Sci. Comput. Program.* **2018**, *167*, 114–137. [CrossRef]
72. Müller, J.P.; Fischer, K. Application Impact of Multi-agent Systems and Technologies: A Survey. In *Agent-Oriented Software Engineering—Reflections on Architectures, Methodologies, Languages, and Frameworks*; Shehory, O., Sturm, A., Eds.; Springer: Berlin/Heidelberg, Germany, 2014; pp. 27–53. [CrossRef]
73. Fasli, M. *Agent Technology for e-Commerce*; Wiley: Hoboken, NJ, USA, 2007.
74. Bergenti, F.; Poggi, A. Multi-agent systems for e-health: Recent projects and initiatives. In Proceedings of the 10th International Workshop on Objects and Agents, Parma, Italy, 9–10 July 2009.
75. Burmeister, B.; Haddadi, A.; Matylis, G. Application of multi-agent systems in traffic and transportation. *IEE Proc. Softw. Eng.* **1997**, *144*, 51–60. [CrossRef]
76. Dudek, G.; Jenkin, M.R.M.; Milios, E.E.; Wilkes, D. A taxonomy for multi-agent robotics. *Auton. Robots* **1996**, *3*, 375–397. [CrossRef]
77. Lee, J.H.; Kim, C.O. Multi-agent systems applications in manufacturing systems and supply chain management: A review paper. *Int. J. Prod. Res.* **2008**, *46*, 233–265. [CrossRef]
78. González-Briones, A.; De La Prieta, F.; Mohamad, M.S.; Omatu, S.; Corchado, J.M. Multi-agent systems applications in energy optimization problems: A state-of-the-art review. *Energies* **2018**, *11*, 1928. [CrossRef]
79. Merabet, G.H.; Essaaidi, M.; Talei, H.; Abid, M.R.; Khalil, N.; Madkour, M.; Benhaddou, D. Applications of Multi-Agent Systems in Smart Grids: A survey. In Proceedings of the 4th International Conference on Multimedia Computing and Systems, ICMCS 2014, Marrakech, Morocco, 14–16 April 2014; pp. 1088–1094. [CrossRef]
80. Zanella, A.; Bui, N.; Castellani, A.P.; Vangelista, L.; Zorzi, M. Internet of Things for Smart Cities. *IEEE Internet Things J.* **2014**, *1*, 22–32. [CrossRef]
81. Brambilla, M.; Ferrante, E.; Birattari, M.; Dorigo, M. Swarm robotics: A review from the swarm engineering perspective. *Swarm Intell.* **2013**, *7*, 1–41. [CrossRef]
82. Ganti, R.K.; Ye, F.; Lei, H. Mobile crowdsensing: Current state and future challenges. *IEEE Commun. Mag.* **2011**, *49*, 32–39. [CrossRef]
83. Bucchiarone, A.; D'Angelo, M.; Pianini, D.; Cabri, G.; De Sanctis, M.; Viroli, M.; Casadei, R.; Dobson, S. On the Social Implications of Collective Adaptive Systems. *IEEE Technol. Soc. Mag.* **2020**, *39*, 36–46. [CrossRef]
84. Annaswamy, A.M.; Malekpour, A.R.; Baros, S. Emerging research topics in control for smart infrastructures. *Annu. Rev. Control* **2016**, *42*, 259–270. [CrossRef]
85. Szuba, T. *Computational Collective Intelligence*; Wiley Series on Parallel and Distributed Computing; Wiley: Hoboken, NJ, USA, 2001.
86. Finkelstein, A.; Kramer, J.; Nuseibeh, B.; Finkelstein, L.; Goedicke, M. Viewpoints: A Framework for Integrating Multiple Perspectives in System Development. *Int. J. Softw. Eng. Knowl. Eng.* **1992**, *2*, 31–57. [CrossRef]

MDPI
St. Alban-Anlage 66
4052 Basel
Switzerland
Tel. +41 61 683 77 34
Fax +41 61 302 89 18
www.mdpi.com

Journal of Sensor and Actuator Networks Editorial Office
E-mail: jsan@mdpi.com
www.mdpi.com/journal/jsan